TALAASH: THE RESEARCH DIARY

INSIGHT PUBLICA®

Nadakkavu, Kozhikode, Kerala, 673011
www.insightpublica.com
e-mail: insightpublica@gmail.com
Title: **TALAASH: THE RESEARCH DIARY**
(English)
Chief Editor : **Dr. Sr. Jaseena Joseph**
Editors: **Dr. Priyadarsini P., Dr. Sangeetha G. Kaimal**
Cover illustration: Sneha K. (Dept.of Physics)
Year of Publication: 2022

Printed and Published by
InsightinPublica Printers & Publishers Pvt. Ltd.
ISBN 978-93-5517-282-2
₹ 259

TALAASH
THE RESEARCH DIARY

Chief Editor
Dr. Sr. Jaseena Joseph

Editors
Dr. Priyadarsini P.
Dr. Sangeetha G. Kaimal

Student Editors
Anasooya Thorakkattu
Rose Maria P. Steephen
Ancy Mariya Alex

INSIGHT PUBLICA®

EDITORIAL BOARD

Chief Editor

Dr. Sr. Jaseena Joseph
Principal
Providence Women's College

Staff Editors

Dr. Priyadarsini P.
Associate Professor & Head
Department of History
Providence Women's College

Dr. Sangeetha G. Kaimal
Assistant Professor
Department of Zoology
Providence Women's College

Student Editors

Anasooya Thorakkattu
B.A. English Literature

Rose Maria P.Steephen
B.A. English Literature

Fida Sherin
B.A. History

Sneha K.
B.Sc.Physics

Sarin Sebastian
B.A. English Literature

Keerthana S.
B.A. Economics

Devika S. Nair
B.A. English Literature

Ancy Mariya Alex
B.Sc.Zoology

Chandana Sunil
B.A. English Literature

Vandana.P
B.Sc.Zoology

Fathima Afra P.K
B.Sc.Computer Science

CONTENTS

MESSAGE FROM THE PRINCIPAL

In the history of Providence Women's College, this is the first attempt by a group of enterprising students to come up with a compilation of thoroughly researched academic papers under the aegis of the Research and Development Cell of the college. They have become pioneers by bringing out this research publication with ISBN – '*Talaash*- The Research Diary' exclusively for students and research scholars. *Talaash*, which literally translates into 'search' is the most ideal word for the title of the publication which encourages research acumen in students as early during their undergraduate days. The Research cell, which was very active during the last academic year, conducted numerous programmes at the national and international level and this diary is its fruitful compilation. The highlight is the earnest attempt made by the cell to give a chance for the students interested in pursuing research to publish their research papers. It is indeed praiseworthy to see the collection of twenty articles from our students and alumnae cutting across various disciplines which stand testimony to their social conscience and scientific temper.

The quality of higher education depends not only on the knowledge it imparts but also the opportunities it provides for the assimilation of diverse skills in students during learning. The editorial board under the indefatigable leadership of Anasooya Thorakkattu & Rose Maria Steephen, Department of English and Ancy Mariya Alex, Department of Zoology have put in their best efforts to compile this work. With much ease and élan they have completed the work on time and I am mightily impressed by the potential and commitment of each student on the editorial board. Being on the editorial board has undoubtedly enriched their student life and I am sure that it would certainly help them to blossom as researchers par excellence in the future.

Credits are due to Dr. Priyadarsini P. and Dr. Sangeetha G. Kaimal for guiding and sharing their expertise with the students in this rare and noble endeavour. *Talaash* will stay close to their heart for all the hard work they have put in to publish it on time.

Let me also place on record the gratitude to our publishers, Insight Publica for the layout and design that has added a lot to the aesthetic quotient of this book.

I ardently hope that *Talaash* continues to be an influence on the research and intellectual life of the campus by catalysing debates and disseminating research output in the coming years. Let this book be an inspiration to many!

Dr. Sr. Jaseena Joseph

EDITORIAL

The Research and Development Cell of Providence Women's College has grown into a highly adaptable and innovative platform over the years, despite the COVID pandemic and the following virtual switchover. The teacher-student coordination and strategic planning and implementation was another aspect that made this academic year more fruitful. *TALAASH: The Research Diary* was born as a result of these challenges and coordination. Promoting a research oriented approach towards learning and knowledge-seeking for students, this book serves to become a noble search for exploring the various dimensions and standpoints on arts, culture, science, law, mathematics and much more. The word *TALAASH* originally comes from the Hindi word तलाश meaning 'pursuit' as a symbolic representation for an active pursuit into the world of knowledge through research and education.

Through this humble initiative, students are receiving an opportunity to delve into their various fields of study with a futuristic vision by incorporating the essence of reality into each of their subjects. The book, being a publication with ISBN, will create a larger exposure and assure a much wider audience for collective thinking and awareness as well as a greater exposure for these budding scholars to let out their ideas.

Like every qualified academic journal, *TALAASH* is also put together with utmost care and critical evaluation taking away any possibility for contingencies. One of the peculiarities of this research diary lies in the fact that the articles, reports, layout, editing and cover were all put together and coordinated chiefly by students themselves. Due to the continued support and encouragement laid out by our mentors and college, this research diary thus enables us to strive towards a highly independent approach to quality research and academic learning.

This project would have been impossible without the planning and hard-work of the Research and Development Cell coordinators and student members. The continued support and efforts from the department faculty coordinators to conduct the programs as well as to collect articles have indeed led this project to completion at a greater pace. Finally, as *TALAASH* unfolds its mastery to all, the pleasure was ours to work in this fine project that is promised to become a great learning platform for students.

Anasooya Thorakkattu,
Rose Maria P. Steephen
Ancy Mariya Alex
Student Editors

MESSAGE

The Research & Development Cell of Providence Women's College was established in 2004 with a strong vision to inculcate research aptitude in young minds and a research publication was a long cherished wish of the cell. At this moment I am indeed excited to see that it has become a reality, finally.

Publications are very important in the lives of researchers as they can define and redefine information and act as a scientific filter. A researcher's performance and productivity is judged largely on the quality of publications and it is certainly the responsibility of higher education institutions to help them pursue their goals.

The launch of this exciting and innovative academic publication will definitely be a milestone in the history of the Research & Development cell and being its executive member since inception, I am really proud of it.

Let me take this opportunity to congratulate the strong team behind this great venture who with their resilience, experience, knowledge, and critical thinking have made it a reality. Their diligence, hard work, late nights, and early mornings have made this a reality. I am left astounded by the dedication and hard work put in by the coordinators of the Research and Development Cell. I wish them the very best in all future endeavours.

May the R&D Cell reach great heights of success. All good wishes!

Dr. Gigy Abraham
Former Head, Dept. of Chemistry
Providence Women's College

OUR ILLUSTRIOUS GUESTS

The R and D cell conducted a national level online workshop on "Resources for Research, Journal Selection and Research Productivity" which was focused in allowing students to have a deeper understanding on research methodologies and research productivities for the students. The program was formally inaugurated by Sri. T. P. Sreenivasan, Former Ambassador and Former Vice Chairman, The Kerala State Higher Education Council. Being in the Indian Foreign Service for 37 years and having nearly 20 years of experience in multilateral diplomacy has led him to become a person with an arduous personality and undoubtedly a highly esteemed presence for the inauguration of the National level online conducted by our college.

The State level online workshop was conducted by the cell on the theme "Introduction to Research and Preparation of Dissertation" for the undergraduate and postgraduate students, and faculty across Kerala. The inaugural session was chaired by Sri. K. Jayakumar IAS (Retd.). He is also the founder Vice Chancellor of Malayalam University. The workshop aimed to instigate research aptitude, benefits of scientific investigation, build awareness on plagiarism, proper methods of academic writing and so on. His accurate knowledge in the sectors of higher education, public services and literature made him the apt fit for starting off such an informative pursuit.

The R and D cell conducted a series of multidisciplinary webinars with a very vibrant faculty group spanning from alumnae members to International scholars. *Khoj*: The International Multidisciplinary Webinar Series was inaugurated with the gracious presence of Shri. P Vijayan IPS who addressed the young minds with his enthusiasm and energy that encouraged the research atmosphere of our college. He was delighted by the idea of the program and gave us plethora of information on research and higher education.

Dr. Rajasi R.S.
Asst. Prof. in Community Medicine
Govt. Medical College, Kozhikode

COMMUNITY ORIENTED ACTIVITIES: COVID VACCINATION AND POST-COVID BLUES

The "development" in "Research and Development" also inculcates the vision of betterment of the community through research. Community development is one among the major objectives of Research and Development. It would be non-fulfilling if the Research and Development Cell of an institution with an extensive reach cannot benefit the community. Acknowledging this reality, the Research and Development Cell of Providence Women's College organised a community oriented webinar on "Covid Vaccination and Post-Covid Blues" - truly a boon for all the attendees, on the inaugural platform of the R & D Cell - 2021-22. The webinar, presided over by Dr. Rajasi R.S, Asst. Prof. in Community Medicine - Govt. Medical College, Kozhikode, was an aid of facts and information on the face of the uncertainty and anxiety which Covid and its vaccination let prevail then.

The webinar was organised by the R & D Cell for the whole college on the 14th of June, 2021 and the Department of Zoology on the 18th of June, 2021 through Google Meet. Dr. Rajasi discussed the importance of getting vaccinated against Covid-19. She explained that vaccines can keep one safe from the dangerous virus and also emphasized on getting 2 doses as it will provide the best, long-lasting protection against

Covid-19. The precautions to be taken before and after the vaccination was also mentioned by the speaker.

The mental stress that people face during Covid times and the ways to overcome them comprised the second part of the webinar, which presented facts and information on the post-covid blues, related symptoms and some tips to regain mental health. Dr. Rajasi reminded us that it was important not to panic and acknowledge that recovery took time. An active Q & A session followed, where several doubts were raised by the participants. The webinar indeed met the community goals of the R & D Cell. It was a zestful 'kick-start' to all the further endeavors of the R & D Cell for the year 2021-22.

KHOJ: INTERNATIONAL MULTIDISCIPLINARY WEBINAR SERIES

The research and development cell of Providence Women's College has had tremendous progress and activities in the year of 2021. Whilst all major programs and events of educational institutions were highly disrupted due to the COVID pandemic, the R and D Cell in association with the 18 departments of the college gave way for one of the most pioneering and informative pursuits for students by kick-starting *KHOJ*. A series of multidisciplinary webinars conducted by premium quality faculty of various disciplines.

The program spanned for twenty-eight days starting from July 23rd 2021 to August 20th 2021. It was graciously inaugurated by Sri. P. Vijayan IPS, Inspector General of Police who described research as a continuous pursuit of knowledge. He also talked about the important skills a student must acquire in the 21st century, like problem solving, critical thinking, communication skills, technical skills and analytical skills while largely appreciating the effort of our college for coming up with this initiative.

Along with it institutional heads of various colleges launched three different programs under the *KHOJ* series.

1. *UDAAN*: The Higher Education Cell was launched by Dr. Godwin Samraj, Principal, Malabar Christian College, Kozhikode

2. *POORV*: An initiative for alumnae in research was launched by Dr. K. Azeez, Principal, PSMO College, Tirurangadi

3. *TALAASH*: The research diary was launched by Dr. B. Rajani, Principal, Zamorin's Guruvayurappan College, Kozhikode

The first webinar was on 24th July 2021, conducted by the Department of Psychology on the topic *Career Perspectives of Psychology in Australia.* The resource person was a Ph.D. scholar of Deakin University, Melbourne, Australia namely, Ms. Sangeetha Thomas. She emphasized the importance of various courses and opportunities around the globe as well as in Australia for Psychology.

On 26th July 2021 webinar under the Department of Mathematics on the topic *Resource Constrained scheduling in Flexible Job shop scheduling problems* took place. It was led by Dr. Jobish Vallikavungal, Professor, School of Mathematics and Physics, University of Nuevo Leon, Mexico.

A webinar on *Qualitative Research in Tourism Studies* was conducted by the Department of Travel and Tourism Management on 29th July 2021 by Dr. Aiwa Romy, Faculty of Social Sciences and Leisure Management from Taylor's University, Malaysia.

The Department of Zoology conducted the next webinar on 30th July, 2021 on the topic *Joys & Tears of Research Where Everything You Dream is on the Other Side of Fear* in which Dr. Nishana M. a research scientist of Langone Medical center, New York University, USA.

On 1st August 2021, the Department of Computer Science organised a webinar on *Research on Computer Science: National and International latitudes,* led by Ms. Deepika Radhakrishnan, Ph.D. scholar, University of New Castle, Australia.

On 2nd August 2021 there were three different webinars under various departments conducted at different timings. Department of Business Administration and Physical Education organised webinars on *Doom Scrolling & Adherence to Physical Activity: A Psychological Approach* respectively. The faculty arranged for the Department of Business Administration were Dr. Pranami Sharma and Dr. Rishikesh K. B. who were assistant professors of Christ University, Bangalore. The Department of Physical Education organised a webinar by Dr. Augustine George, Associate Professor of Physical Education, Government Medical College, Kottayam. Dr. Naveen Nandwana, Associate Professor, Department of Hindi, MLS University, Udaipur, Rajasthan, talked at the webinar organised by the Department of Hindi.

The next webinar took place on 4th August 2021 conducted by the Department of English on the topic *Appropriating Literary Theory* which was delivered by Dr. Priya K. Nair, the Research Officer of Kerala State Higher Education Council and Assistant Professor at St. Teresa's College, Ernakulam. On the same day the Department of Economics conducted a webinar on *Research on International Migration* which by Prof. S. Irudaya Rajan, Chairman of IIMAD, Kerala.

On 6th August 2021 the Departments of Chemistry, Botany and History conducted webinars on *Nanomaterials for Future Energy, Emerging trends for the Young Innovative Minds in Life Science Research* and *Historical Narrative: Science, Modernity and Conceptualization of the Past* respectively. The resource persons for each webinar in the given order were Dr. Subila K. B., Post Doctoral Research Scholar, Artificial Photosynthesis Lab, Department of Chemical Engineering, Ben Gurion University of the Negev, Israel; Dr. Sumitha Thushar, Post Doctoral Fellow, Plant Genetic Resource Scientist, International Centre for Biosaline, Agriculture, Dubai, UAE and Dr. Burton Cleetus, Asst. Professor, School of Social Sciences, Jawaharlal Nehru University, New Delhi.

On 8th August 2021 the Department of Malayalam conducted the webinar with Dr. V. Devi Prasad, Head of DESSH, Regional Institute of Education, Manasa Gangothri, Mysore as the resource person on the topic ശരീരഭാഷയുടെ രംഗപാഠം : കഥകളിയെ മുൻനിർത്തി ഒരു ചിന്ത.

On 9th August 2021 the Department of Political Science organised a webinar on *Public Diplomacy and Indian Foreign Policy.* Sri. Muhasin Puthan Purayil, visiting fellow of the Stinson Centre, Washington DC was the resource person. The Department of Physics conducted a webinar on *Research in Astronomy and Astrophysics: Current Trends* by Dr. Koshy George, Research fellow, University Observatory, Ludwig Maximilian University Munich, Germany.

The Department of French conducted the webinar on the topic *Heuristic Thinking: Unlocking the French Word Genius* on 14th August 2021 by Dr. Arunkumar S., Master trainer (FFL) & Entrepreneur President, Association of Indian French Professionals and Researchers, Puducherry.

Finally on 20th August 2021 the Department of Commerce organised a webinar on the topic *Career in Commerce :The Need for a Strategic Planning* by Sri. Habeeb Rehman, Visiting Fellow, Institute of Management Studies, Cromwell UK International Education, Dubai.

List of Resource Persons

1. Ms. Sangeetha Thomas, Ph.D. scholar, Deakin University, Melbourne, Australia
2. Dr. Jobish Vallikavungal, Professor, School of Mathematics and Physics, University of Nuevo Leon, Mexico
3. Dr. Aiwa Romy, Faculty of Social Sciences and Leisure Management, Taylor's University, Malaysia
4. Dr. Nishana M., Research Scientist, Langone Medical Center, New York University, USA
5. Dr. Deepika Radhakrishnan, Ph.D. scholar, University of New Castle, Australia
6. Dr. Pranami Sharma, Assistant Professor, Christ University, Bangalore.
7. Dr. Rishikesh K. B., Assistant Professor, Christ University, Bangalore
8. Dr. Augustine George, Associate Professor of Physical Education, Government Medical College, Kottayam
9. Dr. Naveen Nandwana, Associate Professor, Department of Hindi, MLS University, Udaipur, Rajasthan

10. Dr. Priya K. Nair, Research officer, Kerala State Higher Education council and Assistant Professor at St. Teresa's College, Ernakulam
11. Prof. S. Irudaya Rajan, Chairman of IIMAD, Kerala
12. Dr. Subila K. B., Post Doctoral Research Scholar, Artificial photosynthesis Lab, Department of Chemical Engineering, Ben Gurion University of the Negev, Israel
13. Dr. Sumitha Thushar, Post Doctoral Fellow, Plant Genetic Resource Scientist, International Centre for Bio-saline, Agriculture, Dubai, UAE
14. Dr. Burton Cleetus, Assistant Professor, School of Social Sciences, Jawaharlal Nehru University, New Delhi
15. Dr. V. Devi Prasad, Head of DESSH, Regional Institute of Education, Manasa Gangothri, Mysore
16. Sri. Muhasin Puthan Purayil, Visiting Fellow of the Stinson Centre, Washington DC
17. Dr. Koshy George, Research fellow, University Observatory, Ludwig Maximilian University Munich, Germany.
18. Dr. Arunkumar S., Master trainer (FFL) & Entrepreneur President, Association of Indian French Professionals and Researchers, Puducherry
19. Sri. Habeeb Rehman, Visiting Fellow, Institute of Management Studies, Cromwell UK International Education, Dubai

POORV
ALUMNAE IN RESEARCH

As part of *KHOJ*, the International Multi-Disciplinary Webinar Series which was the collaborative effort of the Research and Development Cell and the eighteen departments of the college, the project *POORV* - 'Alumnae in Research' was successfully launched in August 2021.

POORV intends to showcase the passion, projects, skills, and expertise of our alumni community through interactions and sharing of knowledge that would be greatly beneficial to the students of our college.

The Department of English had the first speaker of *Poorv* on 10th August 2021, on the virtual platform, Google Meet. The resource person was Ms. Aishwarya Sanath who is currently pursuing her Ph.D. at the Department of Humanities and Social Sciences, IIT Kharagpur. The topic of the webinar was 'Indian Partition of 1947: Monuments, Memory and Mnemonic Fiction'. The talk was on the various aspects of memorialisation of Indian Partition. The scholar shared her experiences of collecting ethnographic data at the Partition Museum and Wagah Border at Amritsar. The webinar shed light on the new theoretical and conceptual fields in English Literary studies. She enlightened the participants with new areas and themes in research that was exciting and intriguing. The speaker also shared some tips for generating interest in research and introduced some journal databases that would help students for research and internships in the future.

The second session of *Poorv* was organised jointly by the Department of Mathematics and History on August 20th 2021, on the virtual platform, Google Meet.The resource person was Ms. Lakshmi Prabha who is currently a Junior Research Fellow at the Department of Mass Communication, University of North Bengal, Darjeeling. Her talk was on 'Editorial cartoons: Visuality and Memory' which focused on Editorial cartoons that according to her are a symbolic device meant to communicate and educate. Being a part of everyday newspaper discourse, editorials cartoons have the potential to simplify the news and build it in a visual language that takes cues from the everyday life of readers. A retrospective analysis of these cartoons will help to build our understanding about how cartoonists in particular and society in general perceived news events. Cartoons in that way can function as a historical source as it captures the collective memory of a society.

Various other departments organised similar webinars which were really useful for the students and gave them inspiration to follow the path of the outstanding alumnae of the college. The sessions truly gave them orientation on how to collect data and write dissertations on the topics which they chose.

Event Gallery

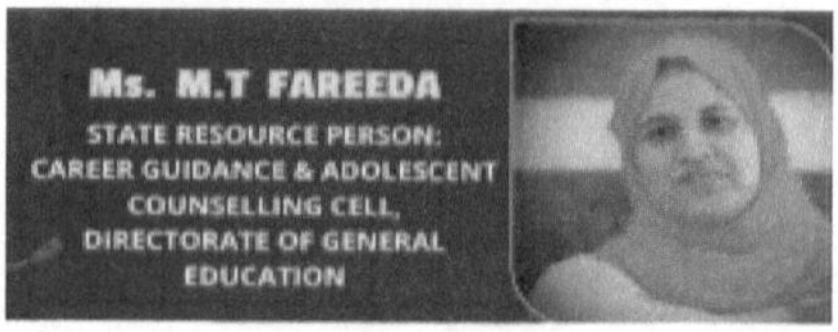

STUDENT SUPPORT PROGRAMMES
AN ORIENTATION WEBINAR ON SCHOLARSHIPS

An orientation on student scholarships, 'You Did Not Rise To Be Mediocre, ' was organized by the Cell in collaboration with the Department of History and the Department of Political Science on 27.06.2021. The program focused on the various scholarship programs offered by the Government of Kerala for undergraduate students.The webinar was inaugurated by Dr. Sr. Ashmitha, Principal, Providence Women's College.The resource person was Ms.M.T. Fareeda, State Resource person of Career Guidance and Adolescent Counseling Cell, Directorate of General Education.Students and faculty from various departments of the college participated in this orientation webinar.

FELICITATION OF DOCTORAL DEGREE AWARDEES

On 25th September 2021, the Research and Development Cell organized a webinar to felicitate Dr. Sr. Asha Thomas, Department of Chemistry, Dr. Sr. Theresia Peter, Department of English and Dr. Nitha Niralda, Department of Mathematics who acquired Ph.D. degree from different prestigious institutions. They spoke about their challenges, struggles and joy with the listeners at the webinar.

Dr. Sr. Asha Thomas's thesis was titled SYNTHETIC AND NATURAL N, S, O DONOR SYSTEMS AS CORROSION INHIBITORS FOR MILD STEEL IN ACID ENVIRONMENTS. The focus of this piece of work is to synthesise and analyse the corrosion inhibition properties of natural as well as synthetic organic compounds containing N, S, O heteroatoms, aromatic rings and π electron systems. We have selected a few pyrazole derivatives and two fruit rind extracts for the study. The fruit rind extract contains a large amount of heterocyclic aromatic compounds which can efficiently reduce corrosion rate. We have synthesised four pyrazole derivatives and the characterisation of the compounds is done using UV, IR, and NMR spectroscopy. The corrosion inhibition property of these molecules towards mild steel corrosion in dilute HCl and H_2SO_4 iscarried out using weight loss, EIS (Electrochemical Impedance Spectroscopy) and PDP (Potentiodynamic

Polarisation) techniques. The Nyquist plots for all the inhibitors have a single capacitive loop which indicates the charge transfer controlled process for corrosion inhibition. Therefore the electrical equivalent circuit used is the Randles circuit. The shift of E corr of the inhibited solutions from the blank solutions are less than ±85mV and which affirms the mixed type behaviour of these molecules affecting both anodic reaction of metal dissolution and cathodic reaction of hydrogen evolution.But few of them show more inclination towards the anodic reaction. The effect of temperature, acid concentration, and inhibitor concentrations on the corrosion inhibition property of the inhibitors are done. Maximum efficiency of 93% is obtained for 100ppm CMPPC in 1M HCl at room temperature. The inhibition efficiency of 89.01%, 90.25%, 84.33% for 200ppm of PD, PD_1 and PY_1 is given in 1M HCl at room temperature. The efficiency is decreased by the increase of temperature and the acid concentrations for all inhibitors. The addition of 50ppm of TU increased the efficiency of PY_1 to 92.36% at room temperature and from 58.31% to 83.41% at 318K. Studies carried out in 0.5M H_2SO_4 give maximum efficiency by CMPPC. Several factors such as molecular structure, solubility, anchoring group etc affect the inhibition capacity of the inhibitors. GIW extract contains o-hydroxy citric acid and cyanidin anthocyanins which could be easily adsorbed on the metal surface. An efficiency of 87.42% at 303K and 80.54% at 333K is given by 4% GIW. The addition of 200ppm of KI to GIW increases the efficiency to 89.05% at 303K. The preadsorbed iodide ions act as a link between the metal surface and cyanidin molecules and thus enhance the inhibition capacity. The ethanolic extract of the fruit rind of Myristica Fragrans (MFE) shows an inhibition efficiency of 96.86% whereas the aqueous extract gives an efficiency of 92.10%. Even at higher temperatures, MFE sustains high efficiency. All the inhibitors show Langmuir adsorption isotherm as the best fitting isotherm and the values of free energy of adsorption (ΔG^0)

indicate the spontaneity of adsorption. The ΔG^0 values for all the inhibitors are below - 40kJ/mol indicating mixed type adsorption involving both physisorption and chemisorptions. The kinetic parameters such as activation energy, enthalpy of activation and entropy of activation show decrease in the rate of metal corrosion by the addition of inhibitor. Geometry optimisation of all the inhibitor molecules is carried out using Gaussian 09 package taking B3LYP 6311G (d, p) basis set. The global descriptors calculated gave an idea about the inhibition property of the molecules. The Fukui indices calculated using the UCA_Fukui software explain the nucleophilic and the electrophilic centres of the inhibitor molecules. The material studio simulations helped in the visualisation of the adsorption pattern of the inhibitor molecules on the metal surface and calculated values of binding energy compared to the adsorption capacity of the molecules. The surface morphology of the blank and the inhibited solutions is observed using SEM and AFM. The roughness of the metal surface is decreased by the adsorption of the inhibitor molecules on the metal surface. SEM images gave an idea of the surface damage caused by the immersion of mild steel samples in an acid medium.

Dr. Sr. Theresia Peter's thesis entitled ENUNCIATION OF THE IGNORED REALITIES IN THE SELECT PLAYS OF MAHESH DATTANI is acquainted with certain social issues in society as expressed in the select plays of Mahesh Dattani. It becomes more familiar with the happenings of the period and proposes to look into the individual life of the characters related to the social life as examined in these plays of Dattani. Man is an integral part of society and consciously or unconsciously the creeping influences of social setup determine the fabric of relationships around him. Instead of dealing with social problems in a traditional way, Dattani represents the unconventional, radical and rational dynamics of social issues, existing in the seeds of the individual self against the predetermined schemes of society. Dattani's plays, *Seven Steps Around the Fire, On a Muggy Night in Mumbai, Tara, Thirty Days in September, Dance Like a Man* and *Ek Alag Mausam* are coloured with real life situations and experiences related with urban cosmopolitan and middle class Indians. The plays selected for this research have made bold innovations and experiments with themes related to the present social scenario, in an effort to bring out a social change and to reform the present social structure. Homosexuality, gender discrimination, life of transgenders, patriarchal setup, incest and love of life are the different themes that Dattani includes in his theatrical work. Therefore this study exposes a realization of life and proposes to question gender roles so as to bring the marginalized identities to mainstream society. In order to formalize the research topic, a review of literature has been included in the introductory chapter of the thesis. The chapter offers a brief look at the playwright Mahesh Dattani and his select plays. It also attempts to show the path of Indian English dramatists in general and distinguishes Dattani from others. It also clarifies the methodology, the objectives and the scope of the present research. His expertise in various theatrical techniques, the verdict of Supreme Court judgment on LGBT people and his interviews are closely scrutinized. Chapter two unravels the interconnectedness between patriarchy and gender – because one cannot be described without the other. It includes a brief assessment of Dattani's plays to find out the elements of patriarchy and gender. This chapter addresses how patriarchy and gender play an important role in the social and cultural existence of the people who are diverted from the mainstream society in terms of sexual preference. It moves on to express that gender victims need space to articulate their position and that can be achieved through their presence in the public domain i.e. in the mainstream. It differs from place to place and person to person in his play after play. Chapter three analyses the alternate sexualities presented in the select plays of Dattani. These plays depict the plight of gays, lesbians and transgenders. Seven Steps Around the Fire speaks about the life of transgenders and their suffering and On a Muggy Night in Mumbai shows the closet experience and the conduct of each gay. Dattani has been advocating on behalf of these deviants as he argues that they should be provided with the same social status as the so-called normal people. This chapter throws light on this gender issue in order to make people think sympathetically and respect everyone's uniqueness. Chapter four appreciates the other themes in the chosen plays. Each of Dattani's plays deals with multiple issues. Therefore this chapter discusses these different issues at length. Themes like family relationships, incest, tradition vs. modernity, HIV, disability, etc. It focuses not only on the major characters but also on the issues related to the secondary/minor characters. Thus this chapter highlights various societal issues that need immediate attention. The concluding chapter sums up the author's presentation of contemporary India highlighting the different issues through his select plays. His characters are life-like and they speak to the audience in their own situations. How the individual faces patriarchy and gender discrimination in the society, the acceptance of the alternate sexualities, and other issues in the select plays are discussed to give an insight into the real life of the people. It also resolves the scope for further studies.

Dr. Nitha Niralda's thesis was titled BAKER MAP AND ITS DYNAMICAL PROPERTIES. The characteristics of dynamical boundaries which are extremely sensitive to initial conditions are explained by chaos theory. Chaotic systems are mathematically deterministic yet erratic due to their sensitivity to initial conditions. Because of the topological transitivity, it is impossible to decompose chaotic maps into two subsystems which do not intersect under the map. Despite all this random behaviour, chaotic systems show a regularity element called periodic points that are dense. The study of dynamical systems and mostly the study of chaotic systems are important breakthroughs of current mathematical research. Baker map, that is a simple two dimensional map, is useful for describing many of the concepts for characterizing chaos. This paper has focused on some dynamical behaviour of one dimensional baker map and has pointed out the MATLAB implementations of dynamical behaviours such as fixed points and their stability, period orbits, and cobweb plot. Also investigated the sensitive dependence on initial conditions and proved the chaotic behaviour of baker map.

CATAPULT: PROPELLING YOUNG MINDS FORWARD

Catapult was an online academic programme launched by *Udaan*, the higher education wing of the Research and Development Cell of Providence Women's College.It was a mega event that comprised almost eighteen interactive sessions with a bunch of our illustrious alumnae and a few successful researchers.The aim of the program was to give students an outlook of the Research process and make them aware of the wide variety of opportunities available for building their carriers.

The main objective of the programme was to guide and orient the students about the cutting edge research facilities and higher education opportunities.The programme comprised of chain of webinars conducted across various disciplines.The webinars were chaired by eminent faculty and alumnae from different parts of the world.The project focused on enriching the students with academic knowledge and research opportunities.

Catapult was kickstarted by a successful and illuminating webinar organized by the Department of Human Resource Management.The session was guided by Mr. Johnsen Clement Madathil, Research scholar in finance and and marketing and Ms. Tejoswini Patra, Research scholar in Human Resource Management, from NIT Calicut.Through the session they shed light on the importance of research, the process of publishing journals and shared own research journeys.Mr. Johnsen

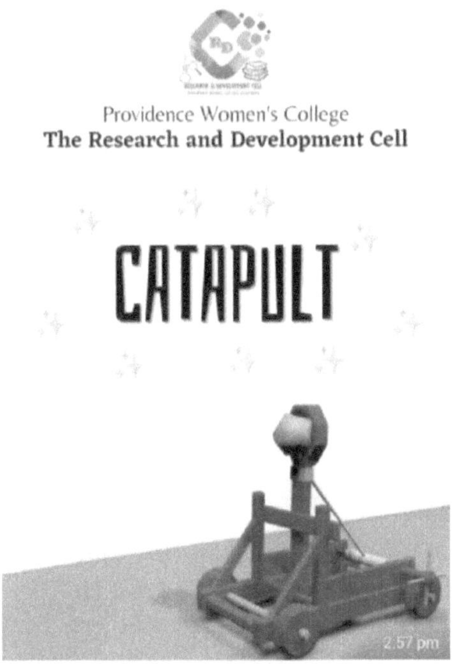

helped the students to understand the educational and technical side of research and Ms.Tejoswini explained to the students the myriad opportunities of HRM and its research side.

Departments of Science also conducted their share of webinars associated with Catapult.The keynote speakers were, Dr. Sheeba P. S. (Associate Professor Department of Chemistry Malabar Christian College, Calicut), Ms. Geethika B. R. (Research scholar, The Institute of Plasma Research, Gujarat), Ms. Gopika G. B (M.Sc. Psychology student, Banaras Hindu University, Uttar Pradesh), Ms. Devapria (M.Phil. Scholar, University of Calicut), Dr. Suba Babu Jayaprakash (Alumna, Department of Zoology), Ms. Dona Alex (Ph.D. Scholar Mathematics, NIT Trichy), Ms. Gopika P. B. (M.Sc. Student Mathematics, NIT Calicut) and Ms. Sreelakshmi P. (Masters in Cyber Security, Northumbria University, UK). All the speakers elucidated the significance of scientific study and the opportunities available both in India and foreign countries in respective disciplines.

Another set of two webinars were arranged by the Department of Commerce. Ms. Swathi Venugopal (M.Sc. in Banking and Finance, Queen Mary University, London, MBA, Heriot Watt University, Scotland) and Ms. Dilna Devadas (Lieutenant Commander, Indian Navy) led the sessions. Ms. Swathi elucidated the advanced opportunities in the field of commerce and shared her experience of overseas study. Ms. Dilna explained her journey and experiences as a Lieutenant Commander. Both the sessions were extremely informative and beneficial for the students.

The webinars organized by the Departments of Humanities and Literature took the whole project to the next level. Ms. Shivangi Shanker (Young India Fellowship Scholar, PG Diploma Liberal Studies, Ashoka University), Ms. Gopika Nair (Alumna, Senior Project Manager -Business Transformation), Ms. Fida Yasmin (M.Phil. Scholar, Modern Indian History, JNU Delhi) and Mr. Gurram Ashok (Ph.D. Scholar, Department of Political Science, University of Hyderabad) were the speakers who led the webinars. They described to students the different ways to explore the possibilities of relative disciplines.The speakers from different parts of the nation also inspired the students to courageously range over their fields of study.

The Departments of Business Administration and Travel and Tourism Management hosted webinars with the resource persons Ms. Athira Das (MBA student, Conestoga College, Canada) and Ms. Aswini Dileep (Travel Agency Manager, Funnel Experts, Dubai) respectively. Both the speakers shared their journeys and explicated different career choices offered by the courses. Students found these sessions valuable and didactic. As part of Catapult, the students and faculty of the Department of Physical Education also explored different carrier options and fields of study along with the resource person Ms. Geeta R. (National Basketball Player, North Eastern Railways, Assam) through a webinar.

Overall, the project Catapult was a grand success. It enlightened the students with vast knowledge and a variety of ideas. It opened the doors of research and higher education. It was yet another fruitful attempt by the Research and Development Cell to make academics more enthusiastic and approachable to the students.

NATIONAL LEVEL WORKSHOP ON RESOURCES FOR RESEARCH, JOURNAL SELECTION AND RESEARCH PRODUCTIVITY

A national level online workshop on "Resources for Research, Journal Selection and Research Productivity" was organized by the Research and Development Cell of Providence Women's College in association with Lore & Ed Research Associates. It was a six day workshop scheduled from 18th December to 23rd December 2021.

The workshop aimed to give a detailed understanding on research methodologies and research productivity to the participants. The session was diligently coordinated by Dr. Priyadarshini P. (Associate Professor & Head, Department of History, Providence Women's College), Dr. Sangeetha G. Kaimal (Assistant Professor, Department of Zoology, Providence Women's College) and Mr. Jinto Michael (Director, Lore & Ed Research Associates). The entire workshop was proficiently conducted via Google Meet.

The whole session in general was formally inaugurated by Sri. T. P. Sreenivasan IFS (Former Ambassador and Former Vice Chairman, The Kerala State Higher Education Council) on 18th December through Google Meet. The inaugural address delivered by Sri. Sreenivasan was enthralling and galvanizing.

The most appealing factor of the workshop was the esteemed panel of experts across the nation who led the session. The panel included, Dr.. Murari P. Tapaswi (Rtd. Chief Scientist (G) (Librarian) CSIR-National Institute of Oceanography, Goa), Dr. I. R. N. Goudar (Adjunct Professor, Department of Library and Information Science, Mangalore University), Dr. Gopakumar V. (Former University Librarian and Head, Department of Library and Information Science, Goa University) and Mr. Viswanath G. S. (Assistant Librarian, School of Planning and Architecture, Vijayawada).

Creating a reading list prior to research, Reference management using Mendeley and Endnote, Document preparation using MS Word, Researcher profile : methods of making research visible, Selecting a journal for research publication, Research productivity measurement based on publications and E-resources for academic research were the major topics covered by the experts.

The workshop was attended by students, faculty and research scholars from various institutions across the country. It was one of its kind and was successful in imparting knowledge and skill to the participants. In general, the session was indeed an informative and worthwhile venture of the R and D Cell.

STATE LEVEL WORKSHOP ON INTRODUCTION TO RESEARCH AND PREPARATION OF DISSERTATION

Research and Development cell of Providence Women's College, Calicut in association with Lore and Ed Research Associates organized an exclusive State Level online workshop on " Introduction to Research and Preparation of Dissertation" for the undergraduate and postgraduate students, and faculty across Kerala. It was a twelve day workshop, which lasted from 2nd February to 26th February 2022 led by dignified panel of experts from all over the state.

The workshop commenced with an inaugural session that was chaired by Sri. K. Jayakumar IAS (Retd.), Director, The Institute of Management in Government. He is also the founder Vice Chancellor of Malayalam University. The keynote speaker of the session was Dr. Johnson Jament (Research Associate, School of Global Studies, University of Sussex, UK). The inaugural session was indeed informative and it did set up the mood for the workshop. The noteworthy research journey of Dr. Johnson was exhilarating enough to attract students to the path of research.

The main objective of the workshop was to introduce the concept of research, its methodology and the writing process, instigate research aptitude among UG/PG students, to raise awareness about the potential benefits of scientific investigation, introduce various forms of scientific writing and the ethical practices involved in academic writing, create awareness about plagiarism, cite sources and manage references.It also provided guidance on the research opportunities in Central and State Universities.

The panel of experts who guided the workshop included Dr. P. N. Harikumar (Professor, Department of Commerce, University of Kerala), Sri. Venkitachalam Sriram (Chief Librarian, R. N. Raj Library, Centre for Development Studies, Trivandrum), Dr. Vimal Kumar V. (Reference Assistant, M.G. University Library, Kottayam), Dr. Vaisakhan Thampi D. S. (Assistant Professor, Department of Physics, Mahatma Gandhi College, Trivandrum), Dr. Praveena K. (Assistant Professor, Department of Economics, NSS College, Cherthala), Dr. Priyadarshini P. (Associate Professor and Head, Department of History, Providence College, Calicut), Sri. Sanjo Jose (Librarian, St. Thomas College, Thrissur), Smt. Lakshmi Dinachandran (Freelance Academic Editor, Physical Sciences and Engineering /Language Consultant), K.M.. Anuja Jose (Research Scholar, IIT Varanasi), Smt. Anjali Mathews (Research Scholar, NIT Calicut), and Dr. Haseena Akbar (Post-Doctoral Fellow, CUSAT, Kochi).

Basic principles of Research, Research methodology, Information literacy, Books and journals, Preparing reading list using Mendeley; Zotero reference management software, Common errors in writing, Report writing and Research and report writing in languages and arts, social science and science disciplines were the major topics covered under the workshop Almost 500 students and faculty from various institutions and Departments across the state attended the workshop.

Recherche-A Digest of Multidisciplinary Explorations, a compilation of 28 research manuscripts encompassing practice-based articles, state-of-the-art content and critical review papers. ***Recherche*** represents the collective thinking of a group of innovative individuals and thus is a rarity in academic publishing. Like all quality academic journals, the book is peer reviewed with rigorous evaluation criteria fully vetted through an editorial board of accomplished scholars representing a wide range of scholarly achievements.

NEP 2020 AND ITS RESEARCH POSSIBILITIES

Devika S Nair & Sarin Sebastian
Department of English
2020-23

Being the first education policy of the 21st century, NEP 2020 aims at the development and progress of our country. This policy revises aspects of the education structure, including its governance and regulation suitable for the current century. This 65 page document, available on the website of the central department of human resource development, if implemented effectively, does have the potential to raise the educational standards up to a global level. This article specifically speaks about the research opportunities NEP offers and how the state of Kerala has reacted to this.

Research, for that matter education, was not a priority in India over the last several decades. Government after government the country only spoke about rural education and research climate with no concrete policy and focus on this subject. According to the World Bank only 0.69% of the GDP is invested in the field of research in our country, in comparison with Germany who spends 3%, USA 2.8 %, China 2%, Japan 3.2%. Out of the 0.69%, 61.4% is utilized for research in defense, atomic energy and space research. According to the World Intellectual Property Organization(WIPO), 70% of the 50, 000 patents are from non-residents. These numerals effectively show how lopsided the research possibilities are in India. According to Preeta George in the article published by Financial Express on 7th December 2020, the key reasons for this abysmal performance are inadequate involvement of industry and the higher education sector, as well as lack of collaboration between these entities and the government in actively promoting R&D. Lack of funding and a low risk appetite, combined with a short-term rather than a long-term vision, are all more or less equal suspects in contributing to R&D insufficiency in India.

PROSPECTS ON RESEARCH

As a game changer to the existing research conditions of our country, the NEP 2020 is introduced. NEP provides a viable research ecosystem under the stewardship of NRF(National Research Fund).One of the goals of NRF is to recognize outstanding research and progress achieved via NRF funding/mentoring across subjects, through prizes and special seminars recognizing the work of the researchers. In most of the committees in the past, the funding agencies engaged members with biased opinions and low knowledge base to the proposals submitted and also as permanent members. By the new NEP, these committees should be more closely linked committees so that it will have an unbiased review process and a strong review input.

Multi disciplinary approach in education, to nurture deep analytical and critical thinking, is another notable proposal included in the policy. It encourages the students to ask questions.By 2040, all higher education institutions (HEIs) shall aim to become multidisciplinary institutions and shall aim to have larger student enrolments preferably in the thousands, for optimal use of infrastructure and resources, and for the creation of vibrant multidisciplinary communities. Therefore, it is a long term vision. Since this process will take time, all HEIs will first plan to become multidisciplinary by 2030, and then gradually increase student strength to the desired levels. NEP also provides an ecosystem with University research institutes run by the government and industries. It provides awards and incentives to encourage students to pursue research. Collaborations between academia and industries and Patent policy structure at University level are the other proposals.

Another encouraging policy change is permitting the researchers to retain IPR (Intellectual Property Right). This is a good incentive and will motivate the researcher to do cutting edge research. Perhaps NEP could have made all faculty positions in HEI as non-tenure track positions so that accountability will automatically fall in line and research output will grow. Then along with non-tenure track policy there should be a strong funding and research support (good research laboratory facility) policy should be in place. Research should go hand in hand with entrepreneurship. This will happen only when along with the basic degree course formulated under NEP 2020, a strong skill development component is included as a part of the academic program.

At higher research levels, all sciences disciplines merge and interaction between these groups is a healthy practice and will help in developing new ideas that may lead to start-ups. NEP 2020 has already been included in the policy creating HEIs in all disciplines. Currently only one of the older IITs has inducted several programs like medicine, economics, management etc. Like in undergraduate programs, graduates in Ph.D. programs should be allowed to transfer credits from one institute to another. In fact at research levels there should be international interaction that will help the student to use laboratory facilities in universities abroad. In order to achieve this, mentors should have collaboration with foreign universities with good laboratory facilities. Best way to achieve this is to have bilateral academies as established by a few IITs.

KERALA'S APPROACH TOWARDS NEP 2020

Initially, Kerala is reluctant to accept the new education policy. It makes several claims as to why the state opposes the policy. Some of them include that the NEP does not clearly state whether this is applicable to all categories of children in all states. If it is implemented as recommended by the policy, hundreds of schools that are suboptimal in terms of economy and operationally complex to run as stated in the report would be shut. Kerala opposes the NEP on the grounds that it is detrimental to the spirit of federalism. Educationists also oppose the new policy claiming that a centralized policy for all states is not a viable idea. For example, the suggestion in the NEP to create clusters in a particular region by closing down smaller schools would only make education non-accessible to many. Dr. Biju-lal, Assistant Professor with Mahatma Gandhi University, told The Federal that such a centralized idea of education, eradicating diversity, is a step towards fascism. He also observes that the NEP is devoid of child rights enshrined in many international covenants. Gender-related themes and provisions across curriculum and the need to recognize gender as a cross-cutting concept is missing in the document. There are also several political reasons behind this negative outlook on the policy.

But now the state has slowly started to warm up to the policy. The State government has launched the Knowledge Economy Mission to bring the ongoing efforts to provide employment to the educated and support 'knowledge workers' under a single program. Higher Education Minister, R. Bindu insisted the government remains firm in its ideological opposition towards the policy due to its scarce regard for social justice, secularism and reservation. She, however, pointed out that the State could not remain ignorant when the country moved forward under a single framework. The students of Kerala would suffer if its universities alone decide against introducing four-year UG degrees. While such an attitude could lead to a large number of students migrating to other States, the State's own graduates might also find scarce job opportunities at a later stage.

The National Education Policy can undeniably bring metamorphic changes to the Indian education system. The long sighted provisions in the field of research are not only beneficial to the research scholars but also will act as a booster for the Indian economy. Though the NEP looks like a game changer in papers, we are skeptical about its practicality. The efforts to make India as a knowledge hub by 2050 is definitely worth a standing ovation. As said earlier, the applicability of the provisions do differ from states to states. The Kerala government has indeed pointed out certain points that the makers should focus on. As we know, Kerala is only one of those states who had closed their gates in front of the NEP. But as pointed out by the Education Minister herself if not implemented with proper changes, Kerala – a knowledge based economy, can forever be in shacks.

LIFE CYCLE OF THE COMMON MORMON BUTTERFLY *PAPILIO POLYTES Linn*

Ancy Mariya Alex
Department of Zoology
2020-23

ABSTRACT

The Common Mormon is a common species of swallowtail butterfly found in Southeast Asia. The life cycle of Common Mormon is completed in the host plants of the family Rutaceae, including *Citrus spp* and *Murraya spp*. The host plant used for the completion of the life cycle in this study was *Murraya koenigii*. Breeding is more active in the months from January to April. The eggs are laid singly on the underside or surface of the leaves or on young stems. The eggs hatched within 3-5 days. Larval stage lasted for 14-16 days, and consisted of 5 instars. Each larval instar lasted for 2-3 days. The Pupal stage lasted for 8-9 days. The development from egg to adult butterfly was completed in about 28-30 days.

INTRODUCTION

The Common Mormon is a member of the Papilionidae family of swallowtail butterflies. It's female is a classic example of Batesian mimicry. Common Mormons can be found in most of the habitats including, tropical or subtropical regions, hilly areas, orchards of the host plant (*Citrus spp*) and even cities. They require access to water or moist earth and plants that are part of the diet of the adults as well as the larvae. The caterpillar eats only new, fully open leaves. Adults feed on nectar of plants including Ixora.

HABITS

Common Mormons cannot fly about under windy conditions. Females are polymorphic and are also typical examples of Batesian mimicry - palatable species mimicking other distasteful butterfly species so as to escape from their predators. The male individuals exhibit puddling behaviour in order to extract minerals, which in turn influence the reproductive success of the species.

DESCRIPTION

Common Mormons are sexually dimorphic swallowtail butterflies with a wingspan of 90-100 mm.The females are generally larger, compared to the males. Male species are black in colour with a row of white spots along the middle of the hind wings and also the margins of the forewings. There may or may not be marginal red crescents on the hindwing, bordering the row of white spots. The upper forewing also bears a row of white spots, decreasing in size towards the apex.

Adult females are polymorphic. *Cyrus*, *stichius*, *romulus* and *theseus* are the polymorphic forms. *Cyrus* form is non-mimetic and is similar to the male, except in having strongly marked red crescents. *Stichius* form mimics the Common Rose butterfly (*Pachliopta aristolochiae*) closely. *Romulus* form mimics the Crimson rose butterfly (*Pachli-*

opta hector). The *theseus* form mimics the black form of the Common Rose butterfly.

The species also contain gynandromorphs that show both male and female characteristics.

HOST PLANT

The Common Mormon requires access to host plants which are part of the diets of both the adult and the larvae. The host plants of the Common Mormon belong to the plant family Rutaceae. They include *Murraya koenigii* (Indian Curry Leaf), *Murraya paniculata* (Orange Jasmine, China Box or Mock Orange,), *Toddalia asiatica*, *Orixa japonica* (East Asian Orixa), *Glycosmis pentaphylla* (Orangeberry or Gin Berry), *Aegle marmelos* (Bengal Quince or Golden Apple), *Citrus maxima* (Pomelo), *Citrus aurantifolia* (Lime) etc. and other *Citrus spp*. The caterpillars are folivores.

LIFE CYCLE

Common Mormon breeding is more active in the months from January to April.

<u>EGGS</u>: Pale yellow, globular eggs are laid singly on the underside or surface of the leaves or on young stems, and change colour to light brown as they mature. The eggs hatched within 3-5 days.

<u>LARVAL STAGE</u>: The larval stage lasted for 14-16 days. It consists of 5 instars. Each instar lasts for about 2-3 days. The larva consumes the eggshell before feeding on the host plant leaves. The caterpillar looks almost the same upto the fourth instar - with a light brown body and characteristic white markings, but grows in size. Except for the fifth instar, the larva resembles bird droppings which serves as camouflage. The fifth instar is green in colour with characteristic transverse bands and an eyespot on either side. The larva engages in feeding most of the time.

PUPA: Eventually, the caterpillar stopped feeding and moved to a lower stem to position itself as a pre-pupatory larva. After the contraction of the body and secretion of a certain silky material, a green pupa was formed, suspended horizontally from the stem with silk thread. The pupa can be either green or brownish in colour. The pupa of Common Mormon has a characteristic pair of anterior horn-like projections and a dorsal hump-like projection. The pupal stage lasted for 8-9 days. The pupa partly displayed black colour with white pattern on the day before eclosion. The adult Common Mormon emerged on almost the ninth day.

The emerged butterfly rested for a while, and then flew into the wild.

CONCLUSION

The life cycle of the Common Mormon butterfly on the host plant *Murraya koenigii* was observed and studied. Each stage of the life cycle was analysed and recorded.

Development of the egg and larval instars of Common Mormon

Fourth and fifth (green) larval instars (left). Pre-pupatory larva and green pupa (right)

Eclosion of Common Mormon (left). Adult male (top right) and female (bottom right) individuals

ACKNOWLEDGEMENTS

The author is grateful to Dr. Sangeetha G. Kaimal, Assistant Professor - Department of Zoology, Providence Women's College, Kozhikode under whose guidance and support this study was conducted and Naveena E. of III BSc Zoology, Providence Women's College, Kozhikode for helpful suggestions on improving the manuscript.

REFERENCES

J. B. ATLURI, S. P. VENKATA RAMANA and C. SUBBA REDDI. 2002. Autecology of the common mormon butterfly, *Papilio polytes* (Lepidoptera : Rhopalocera : Papilionidae). *Journal of Environmental Biology* 23 (2): 199-204.

JOHN CLAUD FORTESCUE FRYER. 1914. An investigation by pedigree breeding into the polymorphism of Papilio polytes. Linn. Philosophical Transactions of the Royal Society of London. Series B. Royal Society Publishing 204: 227-254.

K. SHOBANA, K. MURUGAN and A. NARESH KUMAR. 2010. Influence of host plants on feeding, growth and reproduction of Papilio polytes (The common mormon). Journal of Insect Physiology 56 (9): 1065-1070.

NANDNI DURGESH, RAGHUWANSHI ARUN and SHRIVASTAVA VINOY KUMAR. 2011. Autecology of the common mormon butterfly, Papilio polytes (Lepidoptera: Rhopalocera: Papilionidae). Journal of Entomological Research 35 (2): 101-104.

REVATHY V. S and GEORGE MATHEW. 2014. Identity, biology and bionomics of the Common Mormon, Papilio polytes Linnaeus (Lepidoptera: Papilionidae). IOSR Journal of Environmental Science, Toxicology and Food Technology 8 (1) IV: 119-124.

TASUKU KITAMURA and MICHIO IMAFUKU. 2010. Behavioral Batesian Mimicry Involving Intraspecific Polymorphism in the Butterfly *Papilio polytes. Zoological Science* 27 (3): 217-221.

LIFE CYCLE OF THE GREEN PERGESA HAWKMOTH *PERGESA ACTEUS* ON *COLOCASIA ESCULENTA*

Ajanya K. K.
Department of Zoology
2020-23

INTRODUCTION

Pergesa is a monotypic moth genus in the family Sphingidae first described by Francis Walker in 1856. Its only species, *Pergesa acteus*, the green pergesa hawkmoth, was described by Pieter Cramer in 1779. The green Pergesa hawkmoth, *Pergesa acteus* (Cramer, 1779) is distributed in the Asian continent. They are usually found in the host plant *Colocasia esculenta* in abundantly. The life cycle of the moth involved egg stage, larval and pupal stage before attaining adult status. The larval stage consisting of 5 instars of the moth lasted for 14-17 days. The pupal stage extendedfor 17-18 days.

HABITS

The green pergesa hawkmoth *Pergesa acteus* was found distributed in Sri Lanka, India, Bhutan, Burma/Myanmar, Great Nicobar Island, Thailand, eastern and southern China, Taiwan, Japan, Malaysia, IndonesiaJava and Philippines.

DESCRIPTION

Green pergesa hawkmoth is a nocturnal form with a wingspan of 64–80. Head, thorax and abdomen are purplish grey. Vertex of head and a dorso-lateral stripe to thorax and abdomen is green. Forewings are purplish grey with a green oblique central area from below apex to inner margin with some indistinct lines on it. There is an irregular dark outer area with some yellow inside it. Hindwings with anal patch and sub marginal band are pointed. Male and female looks morphologically alike.

HOST PLANT

The larval host plants are mainly belonged to the family Aracaceae. The most common host plant of green pergesa hawkmoth is *Colocasia esculenta*. The caterpillar feed host plant's leaf and stem.

LIFE CYCLE (Figs. 1 – 8)

The breeding was observed in the month of May. The life stages of the moth involved an egg stage, larval and pupal stage before attaining adult stage.

<u>LARVAL STAGE</u>: The larval stage was lasted for 14-17 days. It consisted of 5 instars. The caterpillar is brown coloured long body with larger eyes, oval shaped six paired of green spots and one pair of large blackish eyespot seen in dorso-laterally.

<u>PUPA</u>: After some days, the caterpillar transformed to a non-feeding pupal stage. The pupa was light brown and yellowish in colour, broadly oval in shape with a curving inwards, under the base forming a cup-shaped depression. At the end of 17 days, the pupa became dark coloured. The following day a beautiful green pergesa hawkmoth emerged from the pupa. The total duration of pupal stage was recorded to be 17-18 days.

<u>ADULT</u>: Characterised by the presence of a pale grey dorsal band running longitudinally from the anterior edge of the thorax to the tip of the abdomen. Forewing apex was strongly falcate; upperside with a characteristic diffuse green band running obliquely across from the inner margin to the apex; discal spot a minute but conspicuous black dot. Upperside of thorax and abdomen had a conspicuous pale grey longitudinal band.

CONCLUSION

The life cycle of green pargsa hawkmoth, *Pergesa acteus* on the host plant *Colocasia esculenta* was observed, photographed and recorded.

Explanation of Figures

Figs. 1 – 3 : Larva of *Pergesa acteus*

Figs. 4 – 5 : Pupa of *Pergesa acteus*

Fig. 6 : Emergence of *Pergesa acteus*

Figs. 7 – 8 : Adult *Pergesa acteus*

REFERENCES

^ *"CATE Creating a Taxonomic eScience - Sphingidae". Cate-sphingidae.org.*

^ *Savela, Markku. "Pergesa Walker, 1856". Lepidoptera and Some Other Life Forms. Retrieved November 24, 2018.*

^ *Hampson, G. F. (1892). The Fauna of British India, Including Ceylon and Burma: Moths Volume I. Taylor and Francis – via Biodiversity Heritage Library.*

^ *Leong, T. M. & D'Rozario, V. "Larval Development and Metamorphosis of the Hawkmoth, Pergesa acteus (Cramer) in Singapore (Lepidoptera: Sphingidae: Macroglossinae)" (PDF). National University of Singapore.*

^ *Pittaway, A. R.; Kitching, I. J. (2018). "Pergesa acteus (Cramer, 1779) -- Green pergesa hawkmoth". Sphingidae of the Eastern Palaearctic.*

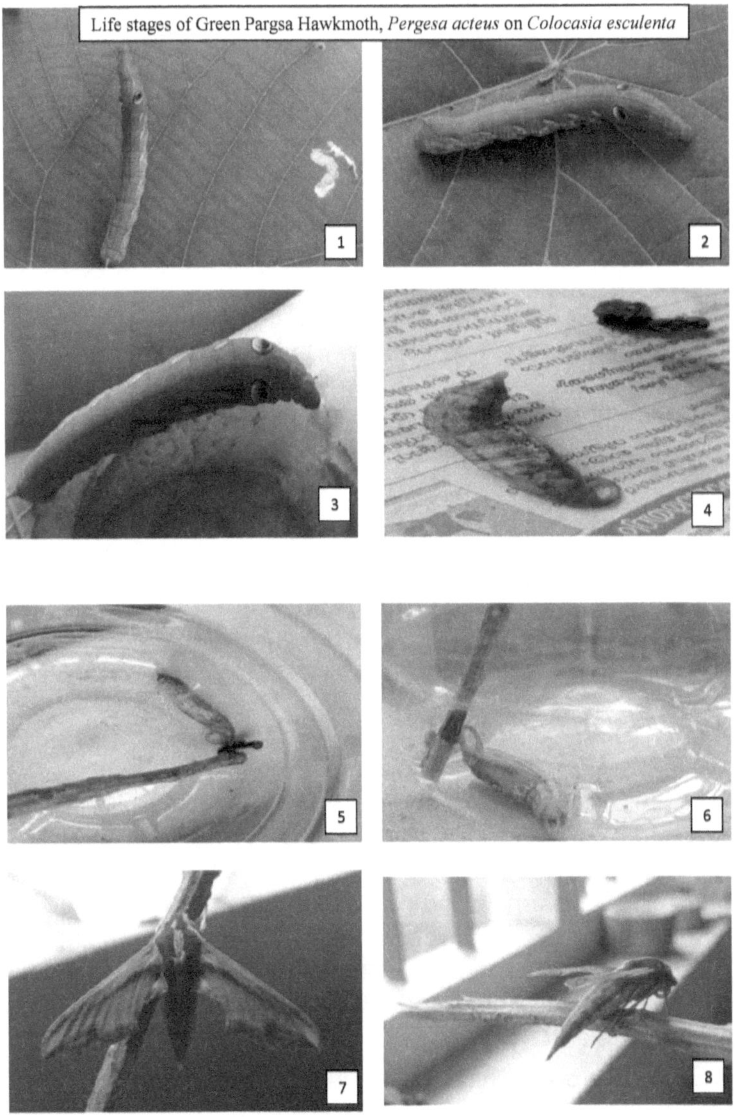

Life stages of Green Pargsa Hawkmoth, *Pergesa acteus* on *Colocasia esculenta*

A STUDY ON CUSTOMER PREFERENCES AND SATISFACTION TOWARDS SELECTED FMCG PRODUCTS OF HINDUSTAN UNILEVER LIMITED

Ambily Benny
Department of Commerce
2019-21

ABSTRACT

Many multinational corporations have made India as their second home over years by accepting the fact that India has the world's second largest population which creates massive number of consumers and marketing opportunity all over the world. Hindustan Unilever Limited (HUL) is a multinational company that has a deep-rooted history in India. This project is a study of the brand preferences and satisfaction of fmcg customers of HUL.

KEYWORDS :- FMCG, HUL, brand preferences.

INTRODUCTION

Customer preference is basically choice of an individual customer, tastes, expectations, likes, dislikes etc measured by permitting the customer to rank several bunches of goods according to their level of utility and selection. It is important to keep track of customer preferences, because both longevity and success of a business directly depend on its customers.

Fast Moving Consumer Goods (FMCG) are product which is popular and frequently used. It refers to products of everyday use. FMCG's are products that are sold quickly and at relatively low cost. It is popularly known as consumer-packaged goods. Items in this category includes all consumables that people buy at a regular interval like toilet soaps, detergents, shampoos, toothpastes, house hold accessories, packaged foodstuff and certain electronic goods. HUL is the most valuable FMCG company. HUL has been one of the most successful in premiumising its brands like Surf, Dove, Lakme and Brooke Bond. According to brand equity, HUL has the largest number of brands in the Top 50, and in the top10.

STATEMENT OF THE PROBLEM

The purpose of the study is to identify the brand preferences and overall satisfaction towards fmcg products of HUL with respect to socio-economic factors particularly in Calicut district.

OBJECTIVES

To identify the brand preferences of customers towards fmcg products of HUL

To evaluate the overall satisfaction of consumers towards fmcg products of HUL with respect to socio-economic factors.

HYPOTHESIS

1. H0 : Satisfaction level on fmcg products and educational qualification of the respondents are independent.
2. H0: Satisfaction level on fmcg products and occupation of the respondents are independent.

RESEARCH METHODOLOGY

This study is based on descriptive research method and it is conducted in various places in Calicut so, convenient sampling is used. The number of items selected for the study were 50 samples. Both primary data and secondary data were used for data collection. Primary data is collected through questionnaire and secondary data are collected through sources like books, journals, magazines, dailies and also from internet. Percentage analysis and Chi-square test were the tools used for data analysis.

ANALYSIS AND INTERPRETATION

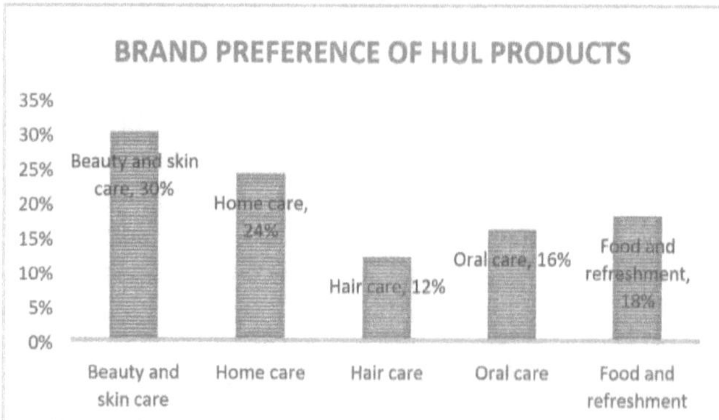

It is clear from the above data that 30 percent of the respondents used the beauty and skin care products, 24 percent used the home care products, 18 percent used the food and refreshment products, 16 percent used the oral care products and 12 percent used hair care products of Hindustan Unilever Ltd.

Relationship between the satisfaction of HUL products and the educational qualification of the respondents.

H_0: Satisfaction on fmcg products and the education is independent.

H_1: Satisfaction on fmcg products and the education is dependent.

Level of significance=0.05

Table value=12.062

Calculated value=7.3701

Since the Table value is greater than calculated value, H_0 is accepted. Therefore, satisfaction on fmcg product is independent on educational qualification. There is no significant difference between the satisfaction level of respondents pursuing degree/diploma, professional, postgraduation and school level qualification.

Relationship between the satisfaction on fmcg products and the occupation of the respondents.

H_0: Satisfaction on fmcg products and the occupation is independent.

H_1: Satisfaction on fmcg products and the occupation is dependent.

Level of significance=0.05

Table value=12.062

Calculated value= 9.1478

Since the Table value is greater than calculated value, H_0 is accepted. Therefore, satisfaction on fmcg products is independent on their occupation. There is no significant difference between the satisfaction on fmcg products of HUL for businessman, professionals, agriculturist and others.

FINDINGS

1. Most of the respondents preferred beauty and skin care products of HUL from all other categories.
2. Majority of the respondents are satisfied with the usage of fmcg products of HUL.
3. All of the respondents are the consumers of Hindustan Unilever Ltd
4. There is no relationship between the satisfaction of customers towards the fmcg products of HUL with respect to the socio-economic factors.

CONCLUSION

In recent years the FMCG sector declined due to down trading. Also, because of presence of large number of companies trying to seize this opportunity, this force the old HUL for to change and thus, their transformation has resulted in a new HUL, which has successfully faced this challenge and reversed this trend. It has done so by substantially strengthening their brands and building capabilities. This has already begun to yield benefits and they are returning to growth. Volume growth is being followed by value growth, which in turn is bringing profit growth.

After studying the consumer preferences, and buying behaviour regarding the various products of Hindustan Unilever Ltd, factors affecting buying behaviour, satisfaction level of consumers etc. are identified. Most of the respondents are the consumers using beauty & skin care products, home care products, oral care products, hair care products and food & refreshment products of HUL.

Today low-priced competition is present in all categories, so without studying consumer behaviour the company cannot survive. And counterfeit products are available in rural areas so company should take steps to stand in the market, and it is their responsibility to make aware their consumers about different products available in the market. Manufacturing of products according to the needs and wants of consumers can help to delight the consumers and it will increase the productivity and profitability of the company.

REFERENCES

Katiyar, A. & Katiyar, N. (2014). An empirical study of Indian consumer buying behaviour of FMCG products (with special reference of bathing soap). *international journal of management and commerce innovations, 2(1), 211-217.*

Ali, M. A., Thumiki, V. R.R., & Khan, N. A.(2012). Factors influencing purchase of FMCG by rural consumers in south India: An empirical study. *International journal of* business research and development. 1(1).

Muthuvelayutham (2012) The Study of Consumer Brand Loyalty on FMCG- Cosmetic Products with Special Reference to Madurai. *European Journal of Scientific Research* ISSN 1450-216X Vol.71 No.1 (2012), pp. 127-143.

Chandrasekhar (2012) Consumer Buying Behaviour and Brand Loyalty in Rural Markets: FMCG. *IOSR Journal of Business and Management (IOSRJBM) ISSN:* 2278-487X Volume 3, Issue 2 (July-Aug. 2012), PP 50-67.

Pallavi, G. S., & Shashidar, S. (2015). A study on consumer behaviour towards selected FMCG. *International journal of engineering and management research (IJEMR), 5(2),* 303-320.

APPLICATION OF NUDGE THEORY IN MULTINATIONAL COMPANIES

Archita Das
Department of Economics
2015-2018

INTRODUCTION

This study attempts to apply the Nudge Theory to the behaviour of the trainees (learners) at a multinational company that is into commodities and finance consulting. The main aim is to identify and correct a negative behaviour of the trainees, without strong-arming them into it. The study was carried out as an experiment on 212 participants, of which 35 were randomly selected to answer a questionnaire. The answers of these 35 participants were then analyzed to understand why these individuals behave in a certain manner; and to deduce any common factors among them. The negative behaviour identified was the trainees showing up late for their classes. This experiment was two months long, of which one month was chosen as the observation period before the nudges were applied, and the next as the observation period after the nudges were applied.

LITERATURE REVIEW

Johnson and Goldstein (2004) in their paper that was published at a time when the shortage of cadaver organ donations was well known and documented. The authors examined the number of organ donors across European countries only to find rather dramatic results. Denmark, Netherlands, the United Kingdom and Germany had really low number of organ donors, juxtaposed to the other seven countries in the study where the numbers were averaging between 80–100%. The extreme difference was explained by the design of the Department of Motor Vehicles (DMV) form. Countries where the forms gave an "opt-in" option had a higher number of donors compared to the four aforementioned countries where the DMV form had an "opt-out" option. In both cases, a large proportion of people left the box unchecked, simply because people preferred to avoid making that difficult and emotional decision of whether or not to donate their organs. This experiment goes on to show how choice-architects and default options have a lot of power in the decisions we make.

Iyengar and Lepper (2000) in Consumers at grocery store encountered a tasting booth. On a regular day, they would find 24 varieties of Jam (extensive) to taste, and on another day at the same grocery store, they would 6 varieties of Jam to taste (limited). The consumers were provided with a coupon to buy the Jams. According to the studies while 60% people approached the booth with the extensive display, and 40% people approached the booth with a limited display, when it came to buying the Jams – 30% people who tasted the limited display bought a jam, while only 3% of the people who tasted the extensive display bought a jam. This study proves that while choices may seem appealing, an overload of choices will tend to generate the wrong result. Choice paralyzes the consumer. The study also stated that the extensive display of jams lead to a lower customer satisfaction than the limited display, proving that too many choices can actually demotivate the customer.

Redelmeier and Eldar Shafir, (1995) was to determine whether situations involving multiple options can paradoxically influence people to choose an option that would have been declined if fewer options were available. Two groups of physicians were mailed a survey containing medical scenarios formulated in one of two versions. A patient that has been treated by the medication has no other option left except to undergo a major surgery after having tried every other medication. The next morning, when the patient was in process before undergoing the surgery, the physician realized that there was one medicine that they could still try out. In such a scenario majority of the physicians responded by saying they would call back their patient and offer the medication instead continuing his process for surgery. The second medical scenario asked to the second group was similar, however, the physicians were provided with one more alternative i.e., there were two medications that they had not tried out. Surprisingly, in this scenario, the majority of the physicians preferred to send their patients for surgery due to the difficulty in deciding between the two medications, which led some physicians to recommend not starting either. In conclusion: the introduction of additional options can increase decision difficulty, and therefore, the tendency to choose a distinctive option or maintain the status quo.

Richard H, (2008) found that in a school cafeteria, students were *nudged* to eat a healthy diet while given the full freedom to choose between healthy food and junk food. In order to indirectly influence their behaviour, the healthy food was placed first or at eye level while the junk food is placed later or in a manner that requires slightly more effort to reach. Additionally, posters promoting a healthy diet were also put up around the cafeteria. The theory of *Loss Aversion* was another chapter in the aforementioned book. Half the students in a class are given coffee mugs with the insignia of their home university embossed on it while the other half of students asked to examine their neighbor's mugs. Then mug owners were invited to sell their mugs and non owners were invited to buy them. They were asked to do so by answering the question "At each of the following prices, indicate whether you would be willing to (give up your mug/buy a mug)." The results show that those with mugs demand roughly twice as much to give up their mugs as others are willing to pay to get one. Thousands of mugs have been used in dozens of replications of this experiment, but the results are nearly always the same. Once an individual has a mug, he or she doesn't want to give it up. However if they don't have one, they don't feel an urgent need to buy one. This implies that people do not assign specific values to objects. When they have to give something up, they are hurt more than they are pleased if they acquire the very same thing, i.e., they are loss averse.

SIGNIFICANCE OF THE STUDY

All organizations aim at being productive in a cost-effective manner. Be it in the case of private firms that run on profit motive or in the case of government (and other charitable) institutions, that run with a social welfare motive – a common factor is being 'cost effective.' This implies not wasting resources; instead utilizing them optimally. It is important to keep in mind that time, effort, and the overall cost of providing a service or creating a product are all costs. This further implies that there is a certain amount that does get wasted when people fail to be punctual. Quite often using 'reverse psychology' can backfire. Salary or wage cuts, for instance, can not only demotivate an employee, but also provoke a negative behaviour pattern. Also, the firm cannot *always* keep offering bonuses; and 'benefits' after a while *will* lose their value due to the law of diminishing marginal utility. Therefore, in order to ensure productivity of the employees, using nudges is an extremely good option as it acts as a positive reinforcement. This study, thus, is extremely significant as it showcases the impact of simple nudges on a firm's efficiency to achieve its goals, without the use of any form of negative or coercive measure. In other words, using positive reinforcement to bring about a desirable change in order to correct a negative behaviour.

Moreover, if the firm can adapt to suit certain needs of the employees i.e., a compromise on behalf of the firm, it would create a sense of belonging among the employees. It is the basic principle of kindness or compassion – when we care about something or someone, we do things for them to make them feel like they're also important. Similarly, employees would not feel insignificant to the organization. This will in turn, motivate them to come to their workplace, and perhaps even engage better. Consequently, there will be an increase in the productivity of the firm leading to better outcomes and positive customer impacts. Nudges, additionally, can also help by maintaining and/or improving the efficiency of employees.

OBJECTIVES

1. To identify the negative behavioral patterns of the trainees coming to class and correct it using a nudge.
2. To find out the reasons for the negative behavioral patterns.
3. To find out the economic impacts of negative behavior.

METHODOLOGY

Study Area: The study area selected to carry out this experiment was a multinational company that focuses on Information Technology related consulting for Finance and Commodities. The organization's learning and development team plans and conducts various learning programs to upskill / reskill / cross-skill the consultants. Upskill is to ensure they learn the newer versions of software and new aspects of functional domains; Re-Skill ensures the consultants learn the newer technology to keep themselves competitive and provide better solutions to their clients as the technology landscape keeps changing at a faster pace. Cross-skill enables people to learn about various products they provide service to their clients and helps in fungibility.

Experiment Period: The total duration of the experiment was two months containing 22 working days each (since, 8 days were weekends in each month). Day 1 to Day 22 was the observation period before the application of the nudges to see how many trainees were late to class. Day 23 to Day 44 was the observation period after the application of the nudges.

Sample Size: 212 trainees were part of the experiment; of which, 35 participants selected for the questionnaire.

Source of Data Collection: After the experiment and observation, 35 respondents were selected and data was collected via direct questionnaire method on the 35 selected respondents.

Tools for Analysis: Measures of central tendencies, pie-charts, histograms, line graphs and bar-graphs were used to analyse the data collected.

LIMITATIONS

The experiment was time and effort consuming as it was carried out over a period of two months. Furthermore, it was difficult to simplify and categorize the various reasons why a person could be delayed. The study was carried out in only one organization; therefore, it is not necessarily accurate as far as other organizations are concerned. Additionally, the answers to the questionnaire are biased according to the participant's perspective i.e., study has not taken into consideration the perspective of the trainer.

SOCIO-ECONOMIC PROFILE OF THE EXPERIMENTERS

The following socio-economic profile provided below is of the 35 trainees who were randomly selected to answer the questionnaire:

Age Bracket: The age brackets the trainees belong to is between 20 years to 32 years of age. It was noted that most of the trainees are 21 years old. The mean age of the trainees is 25.

Gender: Among the trainees, 23 were males, 11 were females and 1 was transgender.

Marital Status: 29 trainees were single; 5 were married and 1 was engaged. Of the 5 who're married, 2 of them have a child each.

Annual Income: Majority earned less than 10 lacs, while the remaining earn between 10 and 20 Lacs, per annum.

Religion: The majority of the trainees were Hindus. While 4 belonged to other religions. There were 31 Hindus, 1 Christian, 1 Muslim, 1 Sikh, 1 Jain.

Years of Experience: There were 17 Freshers i.e. 17 trainees had zero or less than a year's experience. On the other hand, 18 trainees had experience ranging from 3 years to 8 years.

Native States: The trainees work at the Gurugram and Bangalore offices of the firm, but they are natives of different states.

Table 3.1.7. Native States of the Trainees

State	No	State	No
Andrapradesh	2	Bihar	2
Delhi, NCR	5	Jammu & Kashmir	1
Jharkand	2	Karnataka	7
Maharashtra	1	Orissa	2
Tamil Nadu	3	Uttar Pradesh	9
West Bengal	1		

8. Educational Qualification: Majority of the participants have a B.Tech degree, some are pursuing B.Tech and MCA degree.

TYPES OF NUDGES APPLIED

There were three nudges applied that can be categorized as follows –

1. Convenience
2. Motivation
3. Opportunity

After the first month, the trainees were allowed to bring their breakfast to the classroom. In other words, they could eat while attending the class. This convenience was provided to make sure they felt comfortable and didn't have to be in a hurry to finish their food. Additionally, the pressure on a person to run back to class would reduce; and, if they lost track of time while eating and socializing, that factor could also be avoided.

The trainees were also shown five-minute motivational videos before the classes began, two days a week. These videos were **not** related to the subject of training and were taken from YouTube. A couple of videos depicted the 'ripple of kindness' i.e., how one action that an individual may do will create a ripple effect, influencing others to be kind and generous. There were also a few animation videos that comically expressed 'basic class room etiquette, the fact that 'pre-conceived notions or judgments can lead to ill-conceived behaviour' and 'perception'. A handful of videos were the experiences of well-known members of the society – short speeches – where they talked about their journey, inspiring people to not give up when things get tough, to be consistent, to grasp at opportunities when they appear, etc.

The third nudge was an opportunity given to the trainees to present an extempore on the "topic of the week" randomly selected. These topics were related to their work and training subjects.

They found it very exciting since when they spoke, very senior level managers (Directors/Vice-Presidents) were present to listen to what they had to say. The third nudge was an opportunity not easily availed, especially for freshers.

DATA ANALYSIS

The data was collected via survey and via observation of the attendance of the trainees. The survey-data was collected via direct questionnaire method. 35 trainees were selected from the total of 212 trainees. These 35 trainees were selected to answer the questionnaire as they were the ones found coming late to the classes. The main agenda was to find out whether there were any common factors among the people coming late to the classes. For this purpose, general reasons were broadly categorized as Personal, Lack of Infrastructure, Health, Traffic and Others (if any). This segment of the questionnaire was a free-choice question; implying that they were free to select more than one reason. Moreover, the reasons were then expanded into sets of specific reasons (again, for free-selection). Thus, for example, if an individual had selected "Personal Reasons" and "Traffic" as their reason for being late, they had to answer the questions under "Personal Reasons" and "Traffic" – so, the remaining three parts of questions would not apply to them.

It was concluded from the analysis of all the data collected that the 35 late-comers out of the total 212 trainees were coming on time to the classes after the nudges were applied. Additionally, the analysis of the data from the responses to the questionnaire clearly shows that the major reason that the trainees said they got delayed was traffic. A way to interpret this can be that the trainees made a little extra effort by either leaving slightly early or making a slight change in their routine in order to be punctual to class. However, it had been observed that the trainees got delayed further to classes because they often rushed to the cafeteria/canteen and ended up spending more time there due to 'socializing'. Human beings are social beings. Once we start socializing it is easy to lose track of time. This was, of course, evidently corrected by the nudge of convenience. Hence, we can say that the nudges applied showed a positive impact on the behaviour of the trainees.

Looking at the economical impacts of coming late, the costs that the company undertook to provide the 212 trainees with the training was considered. The ballpark value of providing the 3-month long (66 working days) training to 212 people is ₹27, 984, 000. For each working day, the training was 8 hours long. The total loss incurred as a result of 35 trainees coming late to class over a period of 22 days was calculated. The 35 respondents were coming late to class at an average of about 10 minutes. The sum of the number of times all 35 trainees were late is 192. Therefore, the following deductions could be made –

The cost of one person being late by 10 minutes = ₹41.6.

The cost of all 35 trainees being late at an average of 10 minutes over a period of 22 days = 41.6 x 192 = ₹8, 000

Now, if the 35 trainees were late for 3 months, it would cost the firm ₹24, 000. However, since the nudges ensured their punctuality, the amount of loss that was avoided by the firm was ₹16, 000.

Thus, the experiment of using nudges to correct the negative behaviour of the trainees was successful; and, it has benefit the firm by helping it to avoid a loss.

FINDINGS

The experiment was carried out with the objective of identifying and correcting a negative behaviour using nudges. Based on the observations, 35 out of 212 trainees were found to come late. These late-comers were then given a questionnaire. The data collected was then analyzed using statistical tools. The findings and suggestions below have been categorized into two, i.e., based on the perception of the trainees which was derived from their answers from the questionnaire, and based on the reality which was the data derived from observations.

The nudges were successful as the 35 trainees who were found to come late were always punctual to class after the implementation of the nudges.

Based on the Perception of the Trainees

1. The main cause of delay is traffic, as voted for by the majority of the respondents. The second cause were personal reasons.
2. When asked their perception on how often they were late, most of the respondents said they were only sometimes late while a few said they were either never late or mostly late.
3. The office cab services are **not** a cause of delay.
4. Most of the respondents stated that being 10-20 minutes late was tolerable, while four respondents stated that being late was never tolerable.

Based on the Reality

It was observed that the respondents would get delayed mostly because they'd rush to the cafeteria after their arrival and 'socialize', thus getting slightly delayed to the classes.

It was also noticed that most of the participants were either sometimes or mostly late.

SUGGESTIONS

Based on the Trainees' answers

According to the trainees, change in work routines and more creative projects will act as an incentive for them to be more punctual.

Based on the Reality

1. There should be better time management on behalf of the trainees.
2. Since (under personal reasons for delay) searching for wallets, keys, etc is the leading cause for delay, they try and keep their articles at a specific place.
3. The company should provide the employees with more opportunities.
4. The projects should be more creatively challenging.
5. Incentives could be provided so that the employees use the office cab services more often as they run on time.
6. Stress managing techniques can be adopted by the company. For example, an open library for the employees; discounted membership for yoga classes, self-defence classes, stress-management classes, gym etc; a games room in the building with table-games like table-tennis, 8-ball pool etc.

CONCLUSIONS

Based on the observations, the negative behavioral patterns identified were 'not being punctual' or 'coming late to class'. This was corrected by executing three nudges i.e. the convenience of having food while attending the class, the motivation from five-to-ten minutes videos, the opportunity to present an extempore on the 'topic of the week' in the presence of very senior leaders in the organization. Thus, the primary objective was satisfied.The second object was the deduction of the causes that lead to the lateness of the trainees. Two main causes identified were traffic (according to the respondents) and socializing in the cafeteria (based on observation). Besides these causes, according to the respondents searching for wallets, keys, etc under personal reasons was the second highest cause for their delay. Stress and Insomnia (under health reasons) were high impact causes, so the trainees should make sure they take appropriate measure to manage their health – physical and mental – while the company should try to reduce the stress at work, wherever and whenever possible. The third objective was also satisfied as according to the calculations, total duration of training was three months (of approximately 66 working days) and the per day, per person value of the training was evaluated at ₹2000. The total amount of potential loss thus, amounted to ₹24, 000. Since the nudges were applied after the first month i.e. 22 days, based on the data collected from observation, it was calculated that the total amount of loss that was later avoided by the firm was ₹16, 000. Therefore, the economic impacts of the negative behaviour were also found, and loss was avoided, consequently satisfying the third and final objective.

REFERENCES

Defaults and Donation Decisions by Eric J. Johnson and Daniel G. Goldstein, SSRN Transplantation, Vol. 78, No. 12, pp. 1713-1716, 2004

Personality Processes and Individual Differences (The Jam Study) by Sheena S. Iyengar and Mark R. Lepper (2000)

Medical Decision Making In Situations That Offer Multiple Alternatives, Donald A. Redelmeier and Eldar Shafir, JAMA 1995

Nudge: Improving Decisions About Health, Wealth and Happiness, by Richard H. Thaler and Cass R. Sunstein, Yale University Press 2008

YAKSHI AS FEMME FATALE: POLITICS OF THE MONSTROUS FEMININE

Nayanthara Thilak
Department of English
2019-2022

INTRODUCTION

The fear of female sexuality was there from time immemorial, as evidenced in the barbaric deeds of female genital mutilation and Breast ironing carried out by various societies across the world which are aimed at subduing a woman's sexual powers. The vagina of women contains "so much darkness and secrecy surrounding them — like the Bermuda triangle" says Eve Ensler in *The Vagina Monologues* (1). People are ignorant regarding matters of female sexuality because the female body is wrapped in taboo and Shame. When there is lack of clarity and knowledge, people turn to superstition.

The figure of the "Yakshi" in the folktales of Kerala can be used to understand how the female gender is attributed this sense of monstrousness. The article has borrowed the term "Monstrous Feminine" from the paper "Horror and the Monstrous Feminine: An Imaginary Abjection" by Barbara Creed to allude to this concept. Many feminist critics such as Laura

Mulvey have used psychoanalysis as a political weapon. Even though many discourses on psychoanalysis is predominantly misogynistic, the article has used the theory as a tool to further its arguments. The article will delve deeper into the facets that define the "monstrousness" of the Yakshi. It investigates how the symbol of Yakshi has emerged as a site of reverence, fear and pleasure. The main objective of the article is to evaluate the ambiguous status of the monstrous feminine as a subject and object by making a close reading of the Yakshi stories from Kerala.

CONTENT

"All human societies have a conception of the monstrous feminine, of what it is about woman that is shocking, terrifying, horrific, abject." says Barbara Creed in her paper "Horror and the Monstrous Feminine: An Imaginary Abjection" (1). The monstrous feminine is a symbol widely seen in mythologies all around the world. The term signifies a feminine that is outside the boundaries of the phallocentric society. These narratives can be considered as attempts of the phallocentric society to make sense of the feminine which they perceive as the "other". According to Freud, the Perseus myth is an example of the anxieties of the masculine of being castrated by the "other" or the feminine, a narrative which foregrounds female sexuality as different from male sexuality and thus is unknown, deviant and grounded in monstrousness. Similarities can be drawn between the figure of 'Medusa' who is part of the Perseus myth and the story of 'Kalliyankaattu Neeli'. They started out as two virtuous women in the service of the Patriarchal society. When Medusa served the Gods, Neeli served her husband and was loyal to him. However, we find them both being wronged by the very entities they served. Medusa was raped and cursed by the Gods and Neeli was abused and murdered by her husband. Like the Yakshi, Medusa is painted into a half-human, half-animal monster. She is both feminine and monstrous. While she may be known for her monstrosity, her beauty remains just as dangerous. In a male-centred society, the feminization of monsters serves to demonize women. Kalliyamkattu Neeli like Medusa is the ultimate femme fatale

Be it "Yakshis" of Kerala, "Chudail" of North-Western India, " Lamia" from the middle eastern tradition, 'Nu gui' and "Chan chu" from Chinese folklore, " Medusa", "Arachne" and the infamous "Sphinx" from the Theban cycle of Myths, "Mandurugo" from the Philippines, "Penangalan" from Malaysian folklore, 'Lilith' of the Hebrew tradition or the 'witches' of the western world, every tradition finds a way to incorporate the feminine figure as " abject" or one that is shrouded in mystery that is both challenging and frightening to the patriarchal society.

> Abject things are those which highlight the 'fragility of the law' and which exist on the other side of the border which separates out the living subject from that which threatens its extinction. But abjection is not something of which the subject can ever feel free-it is always there, beckoning the self to take up its place, the place where meaning collapses. (Creed 2)
>
> In Julia Kristeva's *Powers of Horror*, Kristeva terms "Abjection", as that which does not 'respect borders, positions, rules'... which 'disturbs identity, system, order'(4). According to Kristeva, the place of the abject is 'the place where meaning collapses' (4).

The "Abject", in short, is a kind of non-object that lingers in a person's psyche as a consequence of repression. According to Kristeva, the abject refers to the human reaction to a threatened breakdown in meaning caused by the loss of the distinction between subject and object or between self and other. The symbol of the Yakshi is an important text that brings out the notion of feminine as the seducing vampire and the witch. The Yakshi occupies an ambiguous territory for the masculine as she is the one with agency or power over the masculine. She creates a meaning that is beyond the system of meanings known to man. She is not the typical woman who submits to the will of the phallocentric society. She is not docile nor is she pliable. She is the perpetrator, the femme fatale; she has the power to kill and determine the fate of her victims. Julia Kristeva says:

> Any crime, because it draws attention to the fragility of law, is abject, but premeditated crime, cunning murder, hypocritical revenge are even more so because they heighten the display of such fragility (Kristeva 13)

The "monstrous feminine" is incorporated into folktales and myths to enforce the symbolic order like the Yakshi tales of Kerala. These Tales from medieval Kerala brings out the literal and metaphorical merging of caste and gender boundaries. These stories act as a tool in the hands of the patriarchal society to maintain their position as the dominant order and identifies deviancy as part of being "abject" or outside the established order or boundaries. The monstrosity of the Yakshi figure can also be attributed to the assimilation process that was carried out in India. When the native culture and tradition of Kerala was integrated into the "great traditions" of Buddhism, Jainism and Brahmanism, they were reduced to "little traditions". Yakshi stories create distinctions between the 'ideal feminine' and the 'monstrous feminine'. The 'ideal feminine' in the folklore of Kerala belongs to the 'great tradition' of deities. The 'ideal feminine' is often the 'mother goddess' who are spouses of male deities of the Hindu pantheon. Stories depicting a clash between the 'ideal feminine' and the 'monstrous feminine' can be read as a clash between 'great tradition' and 'little tradition'. This binary is very evident in the folktale relating to "Chottanikkara" temple, wherein, the main deity who is a form of Goddess Mahalakshmi defeats a Yakshi who was trying to kill her devotee, Guptan Namboothiri. (Sankunni 101). The Yakshi stands as the representative of primitive mother goddess cultures and fertility cults whom different established religions in India tried to control and assimilate the outsider, who even when she was assimilated and transformed still carried traces of that past. Her "wild" sexuality in different established religions was often reinscribed as part of a benevolent fertility cult. Her portrayal as a monstrous figure has much to do with how these mother goddess cults were outside the scope of domestication which was prescribed for women in those days. It can be argued that the villainization of Yakshi has much to do with what she represents for the straight man. She is a mirror to his inner world. She acts as a gateway to the straight man's Id. The sinful nature of his thoughts however makes him resent her. He wants to indulge in his sexual desires but at the same time, he's guilty of nurturing such thoughts as religions of that time considered it to be sinful to indulge in lust.

It's interesting to note how the victim meets his killer at Pathways.In the story "Koshappalli Namboothiriyum Yakshiyum", The Namboothiri finds the Yakshi on one such crossroad. "Oedipus" the hero of the Greek tragedy "Oedipus Rex" written by Sophocles meets the Sphinx, a mythical creature with the head of a human and the body of a lion with thewings of afalcon on his way to thecity of Thebes.Similar characters such as theScottish "Baobhan-Sith", Malaysian "Langsuir" and Portuguese "Bruxsa" also target lonely and lost travellers at night. The monstrous Feminine in this context exist in an in-between space, a greyarea. The setting in which the monstrous feminine appears is lonely and dark which adds to her mystery. She stands on boundary lines that separates what's known and unknown. The victim man feels threatened by this uncertainty. The setting of the folktale is a reflection of his inner fears. The Monstrous Feminine is outside the phallocentric order. Her status as "abject" is evident here.

The motif of Yakshi brings to light the curiosity of the masculine over the feminine and the fear of the unknown. Yakshi's beauty and her seductive quality entices the masculine but she is also a symbol of death. The motif of Yakshi unites two different anxieties of man into one – death and the feminine. Both Death and the Feminine are two concepts that are shrouded in mystery. It is important to note how the female womb is connected to both creation and destruction. Through cultural displays and myths of women's genitals we have learned to associate the vulva with aspects of sexuality, fertility and spirituality. Intrinsically linked to these cycles is the final stage of life culminating in death.

The term "vagina dentata" was coined around 1900s, although the motif has existed for centuries across cultures.Vagina Dentata is Latin for "toothed vagina". It describes a folktale in which a woman's vagina is said to contain teeth with the associated implication that sexual intercourse might result in castration for the man involved. In North America, The tribes of Ponca-Otoe has the myth of a clever masculine Coyote who outwits a wicked old woman who placed teeth in her daughter's vagina and of another young woman whom she kept prisoner, in order to seduce, kill, and rob young men.

The "mantravadi"or the exorcist of the Yakshi stories does exactly this.They bring the deviant Yakshi into the symbolic realm and under the control of the masculine thereby maintaining the traditional systems of meaning wherein the masculine is the more powerful one. This particular trope can be seen in the story of Kayamkulam Yakshi or Krishnapuram Yakshi.The sorcerer Kumara Mangalam Namboothiri contains the Yakshi in a "Narayam"or an iron stylus. She was then taken to his illam but is barred from the inner courtyard of the house. The placement of Yakshi outside the "Naalukettu" is a clear indication of her abject status. The Naalukettu in the story can be compared to society itself which rejects the Yakshi who is an embodiment of Feminine power. Yet, she exists on borders or outskirts of the society. In this story, we find the Yakshi begging the sorcerer to not reduce her to a servant. The mantravadi has the right to determine the fate of the all-powerful Yakshi. Her sin is judged by the sorcerer and appropriate punishment is given to her by him. In many stories the sorcerers end up marrying the Yakshi. The marriage between the sorcerer and Yakshi can be seen as efforts by the phallocentric society to bring back the "abject" Feminine into the domestic sphere and into the realm of the symbolic.In many stories, the yakshi would request the exorcist that she be given a space in a nearby temple and she would thereafter be worshipped as adeity in theouter courtyard of the temple. Thus, the sorcerer acts as an agent of the phallocentric order who brings the chaotic feminine into the control of the masculine order.

CONCLUSION

Yakshi represents the unknown possibilities and potential of women.Although the unknown is a ground of fear, it is also a ground of pleasure and excitement.The main goal of the paper was to analyse the figure of "Yakshi"as a site of fear, reverence and pleasure in a patriarchal context. The main argument of the paper is that the monstrous feminine is an entity which occupies neither the 'subject' position or the 'object' position. "Abjection" signifies her ambiguous role in the phallocentric society. It is the society's way of incorporating the mystery of the feminine into the symbolic realm. The monstrous feminine is a narrative that serves to 'contain' the deviation of the feminine.

Though one may assume Yakshi as the one occupying the subject position because of her immense power, it is not so. An agent of the patriarchal society is introduced in the narrative of the monstrous feminine to relegate her to the status of "abject". Through the "mantravadi" or the sorcerer, the patriarchal society re-establishes their power and authority. All in all, the main goal of the paper was to approach the concept of the monstrous feminine with a new lens. The focus of the paper was to study the villainization of the Yakshi figure and to read into the politics of the monstrous feminine

REFERENCES

Creed, Barbara."Horror and the Monstrous Feminine: An-Imaginary Abjection."*Screen* 27 (1986): *Oxford Journals*. Web.19 Nov 2010.

Ensler, Eve."The Vagina Monologues."*Dramatists Play-Service. Inc*, 2009. Print.

Freud, Sigmund.*The Question of Lay Analysis*. New York:W. W. Norton & Company, 1969. Print.

Kinsley, David. "Hindu Goddesses: Visions of the Divine Feminine in the Hindu Religious Tradition." Delhi: Motilal Banarsidas, 1987.Print.

Kristeva, Julia. *Powers of Horror: An Imaginary Abjection*. Trans.Leon S. Roudiez. Columbia University Press, 1982.Print.

Meenu, B. "Anxious Encounters with the (Monstrous) Other: The Yakshi Tales of Medieval Kerala". *Rupkatha Journal on Interdisciplinary Studies in Humanities* (2020): n.pag. Web. 22 Mar.2020.

Sankunni, Kottarathil. *Eithihyamaala*. Kottayam: Current-Books, 1996. Print.

PERSONA: AN ANALYSIS OF THE FOUR SONGS FROM BTS ALBUM MAP OF THE SOUL :7

Chandana Sunil
Department of English
2019-2022

INTRODUCTION

BTS (Korean: 방탄소년단; RR: *Bangtan Sonyeondan)* also known as Bangtan Boys is a south Korean boy band with seven members which was formed in 2010 and had their debut in 2013 under the Big Hit entertainment(now named Hybe). The septet consists of members RM, Jin, Suga, J-Hope, Jimin, V and Jung kook (stage names).

BTS and their songs never failed to gain world wide attention of music lovers, specifically k-pop fans. BTS together with their fan base known as the 'Army' which is one of the largest fandom of the world had reached great heights of achievements including two Grammy nomination. Nevertheless the band had been very observant and keen on their production of music with which they provide some sort of messages to the world. They have always tried to reinforce their significance as artists through their record breaking songs with deep connotations and had proved their importance to those who refuse to accept them. In this article I intend to analyse 4 songs from their Map of the soul:7 album on the basis of the famous Swiss psychiatrist and psychoanalyst CG Jung's concept of Persona. Music and media, in a way can reach out to vast audience than books can and therefore this extension on psychological concept in music will create a sense of consciousness among the listeners which is a need of the hour. This analysis shows how the band is delving deep into psychology.

ANALYSIS

'Intro : Persona', the solo song by RM (Kim Namjoon) starts with an immediate introduction to the whole theme of the song through the line "*who am I'*. The Latin word 'Persona' means 'Mask' (*Collected works of CG Jung* 363*)* and the titular meaning is traversed throughout the song. Being an artist under an agency would have been difficult to cope up with, but still the team managed to reach the pinnacle of success. The answer to his question will require a lifetime. The six members are seen dancing around him in the music video, which represents the different aspects of a single personality. He is the single voice, they are the different facets of a personality. This is a song of longing and struggles for authenticity. One feels that this personality is struggling to say something, to come to terms with himself, with who he is and what he is. He is struggling with the multitude of himself. "I am still not sure if I'm a dog or a pig or what else." These lyrics from the song portray how he is confused about what he is, what his inner self is. "I dreamt of becoming a superhero. Now it feels like I really became one. But as it goes on there is so much blabbering, one says "Run" another says "Stop".

This one says "look at the forest" that one says " look at the wildflower". Sundry voices behests on him to be this, that and immeasurably more which feels like clamour to him, that confuse him, pressurise him. Being an idol, they are bid to toe the line. They are responsible to their fandom, audience and authorities expecting them to fulfil their expectancy. Their onstage faces are just a mask that hide their unfeigned personality, which they are obliged to obscure. The lyrics then move into a litany of how he feels he is not good enough. He is insecure about himself, feels he is not good enough for music, for truth, for a calling, to muse.

Later, he discerns to be fake, hiding behind the mask. " There is something that raised me up again every time". The force that drives him up again every time he feels sham may be the inner him, which is the core of your being. In Jungian psychology this is called the self. The real 'you' from the day you were born and is the ultimate wellspring of the psyche. " My name is R. The 'me' that I remember and people know....... The 'me' that I want myself to be.... The 'me' that's sometimes in tears". He is not camaflogue anymore, the 'persona', the real him is unmasked. He is not embarrassed anymore to show his real identity. The 'self' is transpired. Jung considered that from the birth every individual has an original sense of wholeness -of the self but with the development of a separate ego, consciousness crystallizes out of the original feel of unity. This process of ego-differentiation provided the task of first half of one's life course. Once this ego-differentiation is achieved and the individual is anchored in the external world, a new task arose for the second half of the life, a return to, or conscious rediscovery of the self; Individuation ("Self in Jungian psychology.). Here RM is undergoing the process of individuation, after experiencing hardships and criticism from the external world where he hides himself behind the mask, but now rediscover his self, which is a conscious process.

'Make it Right' is a wonderful song about the peregrination to find the soul. They are trying to come to themselves pulling out of the mask. It allows a person to adapt to the world around them and fit in with the society in which they are living. The song exclaims " I'm singing to find you". The 'You' talked about here can be the soul. They was lost trying to find answers in the world around them but the only one who can understand them, the light that paved way in those struggling roads was their inner self. They are trying to find their real selves hiding behind those different social masks. They became heroes, received lots of prizes, achievements but at the end everything is reaching out to themselves. "It's the answer to my journey, I'm singing to find you". The journey to reach their inner self that have been locked up till now. They might mess up, but only they can make it right. The song seems to be talking to someone, but in reality it can be to their own selves. They tries to find their souls in others and through others. "A little taller than before, With a voice a little more solid. All of this go back to you. Now I will spread a wide map called you, my rehab". Even when they change, everything goes back to themselves, trying to restore themselves. From being a small kpop group, the band conquered heights that they couldn't even imagine. There has been tremendous changes within them, their social status and even with their personality. The eternal night with no end in sight mentioned in the song can be the life ahead of them, which they would never know what would happen. They may try to find their soul which can be inner or outer or something that combines the both. "Uninvited. Unwelcomed. The only one who recognized me" this line evokes the idea that the uninvited, unwelcomed one who is the only one who can recognizes the real them is the soul they are trying to find. It is only them who can hold on to them tighter and make everything right.

'Dionysus', is a song of celebration and breaking out of the persona. Dionysus was a Greek God who came from Thrace and invaded Greek. He tramps down people's resistance, one cant hold out against him when he turns up. Even if one tries to counter him, he will overwhelm you. This is a song of how this persona BTS has gotten into the area of Dionysian celebration. They are no longer frantic of Dionysus. In a way, they can embrace him and can also relish his inebriation. He was named for looseness, he dwindled the old compositions, unalterable deportments and decimate the personas. Even if it is for a moment, he frees one from his personas. "Just get drunk like Dionysus. Drink in one hand, Thyrsus on other". I'll show you, I recommend you something different". They are ready to come out of their persona, which traps them without their knowledge, which turns them into a different personality for the sake of external world or society. The different masks each of them wears to shroud their real self is an unconscious process of their psyche.

Here in this song, they are trying to break out the persona that sets traps and they also celebrates Dionysus. Every human need to explode and come out of their masks in order to regain their self that has been repressed till then. Humans needs this release that Dionysus allows. BTS, through this song proclaims this release to repossess the real selves. They are always behind the mask for the sake of the society, their real self is repressed and have a strong urge to blow up the pseudo self. " I'm now in front of the door to the world. The cheers I hear when I get up on stage, can't you see my stacked broken thyrsus. At last I'm reborn". They are standing in front of the door that lead them to their world of real self, stacked thyrsus symbolises the mistakes they had committed, their broken personas. But, in the fullness of time, they reborn. Their real self is regained, personas are shattered. "Born as a Kpop idol, reborn as an artist. What does it matter if I'm an idol or artist, the new record is the fight against oneself, a fight, yeah". Respective line shows how locked up their life is, born as Kpop idol, but then reborn as an artist. But what does it all matter, life is a fight against themselves, fight to improve, fight to move on, every new record that have been created is a fight against oneself to create another record. Trying to get out of the persona that hides the real you is an incredibly dangerous tasks. If one remains unconscious of his true individuality, even though it serves the purpose of being adapted to a society and culture and a well structured persona helps for the practical purpose of survival and success, this in any way does not help in the process of Individuation. Through the song of the celebration of Dionysus, BTS indicates that they are geared up to let go of the persona and also to get away from the traps persona sets. They arrived at a point of disentanglement from the curtailments of persona.

Jimin's (Park Jimin) solo song 'Filter' comes up with a refrain "Which me do you want?" through out the song. There isn't one persona, but different personas. He can be anything that you asks, he can be different for each one of us. Different personas within a human being. The world doesn't see what lies beyond the mask, it conceal the real you. Persona segregate the conscious ego from the outside world and converse with it. When someone creates an opinion on him, they are not looking inside his psyche or ego, but rather at his persona that he has been visible to the world. His persona varies depending on what role he takes up. This interpret his question, which him do we want. Being an idol he has been provided with a persona and has to fit in it to adapt the societal situations. He also has to create different persona for himself, which fit his specific needs, to express his individual personality.

"Mix the colours in a palette, pick your filter, which me do you want". Life is compared to colours, mixing them creates a entirely different colour, similarly, mixing all those personas one has to fit in, it creates another persona which is absolutely severed from the true identity. Like those filters employed in the phones, which swap the real face with something different, he asks to pick a filter that can swap his persona to something different. "Overcoming different tastes and standards, you will be wanting only me. Yeah, because you are the one who made me". Here he addresses to himself, who overcomes different tastes and standards others offered him, his real self is what he really is because it is his true identity from his time of birth but had been curtained with different personas. Psyche has the power of self-deception and is adapted to it." I'm a brand new filter that you have never seen before. Entrust yourself to me. To see more thrilling things, pick your filter. Contain only me". The process of individuation is accelerated only when the persona starts to disintegrate. The disintegration of persona, where one hides his true self is a risky and dangerous effort. The real self contains only him, no personas, no hidings, only the authentic self is retained. To reach the place where the real self rests, we should entrust ourselves, dismantle the fear of being revealed and cast off the constraints of the persona. When circumstances arise that pressure us to reckon with whom we 'really' are in lieu of who we accepted as our true self or who we desiderate ourselves to be, the unconscious may step forward. Disclosing your authentic self can be extremely minacious. There is a menace of being discarded and disparaged. Any commitment to serving social adaptation carries the peril of betraying the real self. The mask every person wears moulds with each dwelling he constructs. The song suggest the concept of persona and how persona changes with change in the situation. Human life switches personas like filters to adapt to the conditions. Choosing different personas is an unconscious process by the psyche and each persona opens up a different world. The final lyric of the song points out, "we will become a new us. Overlay me in your heart". Each time the persona changes, the personality becomes entirely new, extremely different from the past one.

CONCLUSION

This article does an analysis through the lens of analytical psychology found and propelled by Carl Gustav Jung (which is different from Freud's theory of Psychoanalysis), of the selected tracks from BTS's Map of the Soul: 7 album. The album consists of 19 tracks of which 4 songs are taken for the analysis. Analytical psychology deals with human mind, personality and thought and its inner structure and processes, where archetypes play a key role. It takes into account the individual unconscious and its relationship to the collective unconscious. The 4 songs through their lyrics has brought out the depth at which the band had composed the songs delving into Jungian concept 'Persona'.

The four songs analysed in this article subsumes the concept of Persona. These songs shows that every human being is camaflogue behind a mask, the persona. This mask conceals their real inner self, their core of being. Through this songs the members are seen trying to come out of their persona, trying to find out their self and also the celebrating their breaking out of persona. K-pop songs with it's colourful visuals and captivating tunes are often marginalized as mere entertainment songs. This article aims to understand the significance of these four songs which had a wholesome message packed between these visuals and captivating lyrics and also bring out the psychological significance of these songs.

WORKS CITED

1. Mi-Sung, Ahn. *Jungian Scholar on BTS's Soul Searching Album.* K-pop Herald. 28 May. 2018. Web. 5 June.2021. <http://m.kpopherald.com/view.php?ud=201905281654373745017_2>.
2. Stein, Murray. *Jung's Map of the Soul: An Introduction.* Chicago. Open court. 2006. PDF.
3. "BTS". *Wikipedia.* Wikimedia Foundation, n.d. Web.3 January 2022.

4. "Self in Jungian psychology." *Wikipedia*. Wikimedia Foundation, n.d. Web. 6 August. 2021.
5. Jung, C and Adler. *Collected works of C.G JUNG, Volume 7.* Princeton. Princeton University Press. 2014. Print.
6. BTS " Intro: Persona". Music video. *YouTube*. YouTube. 29 March2019. Web. 6 June 2021.
7. BTS. " Dionysus (English translation)". *Genius.* n.d. Web. 12 June 2021.
8. BTS. "Filter (English translation). *Genius.* n.d. Web. 18 June 2021.
9. BTS. "Make It Right". Music video. *YouTube*. YouTube. 18 October 2019. Web. 14 June 2021.

CANOLLY CANAL: ITS SIGNIFICANCE IN THE PAST AND PRESENT

Arya Govind
Department of History
2019-22

The Canolly Canal is a canal network that was constructed under the orders of the British collector H.V.Conolly at around 1848, to facilitate inland trade. It combines numerous rivers along the coast of Kerala, with the intention of creating a vast waterway from Kozhikode to Kochi. This project aims at discussing the significance of the Canolly canal through the ages. It is divided into three chapters. The first chapter deals with the development of transport in Malabar, the second chapter is about Hendry Valentine Connolly and his contributions and the final chapter is about the Canolly canal.

The first chapter deals with the development of roads, railways and canals in Malabar region. The people of Malabar mainly depended on water ways for their transport until 18th century. Construction of few land routes were undertaken during the Mysorean interlude from 1776-1793.[1] Water transportation

1 P Ibrahim, 'The Development of Transport Facilities in Kerala: A Historical Review', Social Scientist Vol6 No.13, 1978, pp34-35

was common because wheeled traffic and pack bullock traffic were not known to the people of Malabar then[2]. The principle reason for the absence of roads possibly were the availability of cheap waterways adequately to meet the requirement of trade and commerce and also the geographical and climatic condition were unfavourable for construction of roads[3] The first ever canal constructed in the region was the 2.4 mile long Sultan Canal, built by the Ali Raja of Cannanore.[4] After the coming of the British, development of roads were temporaly arrested, because their motive was extraction of surplus revenue and to have monopoly of pepper trade. Roads constructed by the British were primarily to suppress the Pazhassi and Mapilla rebellions. They constructed canals like the Payyoli, Ponnani, Beypore, Canolly and the Vadakara canal.[5] The coming of railways lead to a decline in the use of waterways. The first ever railway line connected Tirur and Beypore and covered a distance of 19 miles. It was commissioned on 12th march 1851[6]

The second chapter deals with Conolly and his contributions. Hendry Valentine Conolly was an able administrator who served as the collector of Malabar from 1841 to 1855. He is famous for the contributions he made to the people of Malabar. The Nilambur Teak plantation also known as Connolly plot is one among his contributions. It was established to meet the deficiency of Teak in the Bombay ship manufacturing company. The land of Trikalur Devaswam was utilised for this purpose. As the Devaswam faced financial issues company helped them and used their land in return[7]. He was known for the concern he had for the people of Malabar. He took

---

2 William logan, Malabar Manual Vol I, Mathrubhumi Books, Kozhikode, 2019, p.8

3 P Ibrahim, op.cit, pp34-48

4 William logan, op.cit., pp10-11

5 *Ibid.*, p 267

6 B A Prakash, 'Economic history of Kerala from 1800 to 1947AD Part I: Malabar', Thiruvanandapuram Economic Studies Society, Thiruvanandapuram, 2018, p. 51

7 https://littleflowercollege.edu.in/upload/e_contents/files/586022c720e388e81c83603b9dfcfc86.pdf

intiative in transforming the thatched roofing of the main market veliyangadi to clay tile roofing, as the area was prone to fire out breaks. Velyangadi was one of the famous and largest market in the region and almost all the shops of the region used palm leaves of coconut leafs for roofing. If it gets on fire, it would lead to a serious disaster. In order to avoid this he granted permission and financial aid for doing the roofing, in a period when only the buildings of temples and Brahmins were allowed to have clay tile roofing.[8] He was also known for his role as a social reformer. According to Selluraj, a notable lawyer and writer, Conolly emancipated the dalits much before the Vaykkom Sathyagraha[9]. He had asked the Dalits to use public roads at a time when they were not allowed to use. He also issued orders to stop slave trade in 1843.[10] One of his most important contribution was the Canolly canal

The last module deals with the Canolly canal. The Canolly canal is an integral part of Malabar region and was built only by using man power in 1848. The canal was used mainly for trade in the 19th century. Canolly canal is a part of the west coast canal network that connect Korapuzha River in the north to Kallai River in the south.[11] The width of the canal is 6 meter to 20 meter and the depth range from 0.5 to 2 meters[12]. The canal was used for easy transport of products from hinterlands to Kozhikode town and iron implements from the town back to the hinterlands. The canal construction needed lot of effort. A serious issue was that Land Acquisition act was not passed then. But the ruler of Kozhikode -Samuthiri offered all his help to this. An agreement was even signed with Samoothiri that the canal will be used for transportation and adequate measures will be taken to check salt water uprise to cultiva-

---

8 T B Selluraj, Kozhikodinte Paythrikam, The Mathrubhumi Books, Kozhikode, 2011, p85-86

9 Staff reporter, 'Connolly's contribution remembered', The Hindu, Kozhikode, September 12, 2018

10 The Hindu, Kozhikode, op.cit.,

11 T B Selluraj, op.cit., p127

12 Feasibility report, Irrigation Department Kozhikode, Government of Kerala

ble lands of the region.[13] The workers were given meals as their salary.[14] The canal connected river Chaliyar, Kallai and Kuttiyadi by interlinking these rivers through man made canals[15]. An amount of 12, 416 was spent on the construction of the canal from Ponnani to Chavakadu[16]. In early times water from the canal was even used for cultivation. After the death of Conolly, Major Sally was responsible for the construction of the canal. There were even plans to drop the project and do coconut cultivation in the area acquired because that seemed to be more profitable. But the collector of Malabar Robinson protested against it and the construction was restarted.

Salt water uprise became a major issue that the canal caused according to the Ponnani court verdict of 1890 the canal was constructed for transport and this company was not responsible for the salt water uprise in the region. The development of roads and railways lead to reduction in the importance of the canal. Thus it became neglected over the years. No maintenance work were done by the authorities thereafter. The canal thus became a place to dump waste. Solid waste are disposed-off directly into the canal, especially in city area. Biodegradable waste settles in the bottom of the canal and the synthetic waste floats on the top. Solid waste blocks drains and causes stagnation of water in the canal. Untreated waste from hospitals, industries, factories, and from domestic household are dumped directly into the canal. Domestic waste are comparatively harmless when compared to waste from hospitals and factories.[17]

Government had initiated numerous projects for the development of the canal, but nothing seemed to have worked out. Mr. N. M. Salim had put forward a project in 2013 for the development of the canal and for using it for tourism purpose.[18] In

---

13 T B Selluraj, op.cit., p128

14 *Ibid.*,

15 Feasibility report, Irrigation Department Kozhikode, Government of Kerala

16 T B Selluraj, op.cit., p128

17 Feasibility report, op.cit.,

18 'Conolly canal kendram aaki oru swapna padhathi', Mathrubhumi Nagaram, 6 June 2013

2019, the cleaning of the canal was conducted and proposal for canal maintenance was put forward but presently we see that the canal is full of waste material and the condition of the canal is degrading day-to-day.[19] Mathrubhumi newspaper of February 2022 shows that 180 crore has been passed for the development of the canal[20]. Let us hope that this scheme will be implemented properly. Connolly canal being a resource if used effectively can lead to improving the living condition of the people of Malabar.

BIBLIOGRAPHY

PRIMARY SOURCES

1. B.No.1, 7.2/1856, '*Quantity of tobacco illegally introduced into Angengo under some misunderstanding of late Mr Connolly*', Subject list of revenue record(1858-1890)
2. B.No.2, 5.3/60, '*teak plantation along areas near Beypore river*', Subject list of revenue record(1858-1890)
3. B.No.243, Malabar collector RDis File No. 2954/20, Munsiff court, serial no 26-27, OS 181/1920
4. Census Handbook 1951, Malabar district, p.4
5. '*Conolly canal kendram aki oru swapna padhathi*', Mathrubhumi Nagaram, 6 June 2013
6. '*Conolly canal vikasanathine 118 kodi*', Mathrubhumi Daily, 18 February 2022
7. Disposal No.7/Rev 78, '*Regarding list of roads in Malabar*', July 1878
8. Feasibility Report, Irrigation Department Kozhikode, Government of Kerala
9. Logan William, *Malabar Manual Vol I,* Mathrubhumi Books, Kozhikode, 2019
10. Innes C A, Malabar District Gazetteers 1904, Kozhiko-

---

19 https://www.youtube.com/watch?v=NulwT-kB4FY&t=1s, Boat service and Floating market in Connolly can likely after cleaning drive, August7, 2019

20 'Conolly canal vikasanathine 118 kodi', Mathrubhumi Daily, 18 February 2022

de district, 13/C/24, 1904-05
11. Menon Sreedharan A, *Kerala District Gazetter,* Kozhikode District, 22/G/66, 1962
12. Statistics of Malabar 1873-74

SECONDARY SOURCES

1. Balan C, 'Resource appropriation and development of transport Facilities in kerala during colonial period', *Proceedings of Indian History Congress*, Vol 67(2006-2007)
2. Boase Fredic, 'Modern English Bibliography', Conolly Hendry Valentine(1806-1855)
3. Buckland Edward Charles, 'Dictionary of Indian Biography', Conolly Hendry Valentine(1806-1855)
4. Chincher H U(rev James Hunt) (2004), 'Conolly Arthur (1806-1855)',
5. *Oxford Dictionary of National Biography*, Oxford University Press
6. https://forest.kerala.gov.in/index.php/about-us/forest-dept-history, Keralaforest&wildlife department
7. https://littleflowercollege.edu.in/upload/e_contents/files/586022c 720e388e81c83603b9dfcfc86.pdf
8. https://www.youtube.com/watch?v=NulwT-kB-4FY&t=1s, Boat service and Floating market in Conolly can likely after cleaning drive, August7, 2019
9. Ibrahim P, 'The Development of Transport Facilities in Kerala: A Historical Revive', *Social Scientist Vol6 No.13*, 1978
10. Kareem C K Dr, Kerala under Hyder Ali and Tippu, *State Institute of Language*, Thiruvandapuram, 1987
11. Prakash B A, 'Economic history of kerala from 1800 t 1947AD PartI:Malabar', *Thiruvanandapuram economic studies society*, Thiruvanandapuram, 2018
12. Selluraj T B, Kozhikodinte Paythrikam, The mathrubhumi Books, Kozhikode, 2011

13. Staff Reporter, ‘Conolly contribution remembered’, The Hindu, Kozhikode, 12 January 2018
14. Swai Banavenlore, ‘Notes on the colonial stae with reference to Malabar in 18th and 19th cebtury’, Social Scientist Vol6 No.8, 1978

SIMULATION OF BASIC ATMOSPHERIC PARAMETERS USING PYTHON

Ayisha Fida, Liyana Fathima & Nayanthara
Department of Mathematics
2018-21

Atmosphere and its parameters play a vital role in shaping lives on earth and maintaining it. Our major aim through this project was to study the variations caused by the atmospheric parameters corresponding to a particular altitude. The most important part of our project was to study and use of the programming tool PYTHON. The atmospheric parameters such as density, pressure etc. are functions of height, which is graphed using python. The result to be obtained was the graphical representation of variation of atmospheric parameters. In order to graph all these, it is obvious to learn the programming tool Python, which was a challenging task to us during the Covid era, which was the time of online classes and gatherings. As a part of learning we successfully completed an online course named 'Data Science with Python Course: Hands-on Data science 2021' on January 2021. The project was carried out under the guidance of Dr. Sandhya M, Women Scientist, Department of Physics, Providence Women's College, Calicut. The support and ideas she provided us was our greatest inspiration. Mrs.Aiswarya Paul, Head of Department of Mathematics, Providence Women's College, provided the necessary facilities while the project was undertaken.

Python is a powerful, versatile programming language which has varieties of utilities in many fields which includes the current popular world of data science. This programming language has a simple syntax that is easy to learn. Now-a-days the usage of python language is increasing day by day. In this project we made an attempt to extend the possibilities of python language for the simulation of atmospheric parameters. Mainly atmospheric density, pressure and amplitude are those parameters. The change in these atmospheric parameters with increase in height from the surface of earth is observed. There are mathematical formulas which describes this relation. Using those formulas, the relations can be graphed. A graph is necessary whenever one describes variations of one variable with respect to another variable. With the help of python different sets of values can be taken and graphed. Comparisons are made easier. Anaconda is a computer program which includes Python and R programming languages for scientific computing. After installing Anaconda, it is free to access 'Spyder', 'Jupyter notebook' and many other applications inside it. 'Spyder' was the application used to plot the graphs.

This project is mainly carried out through three ways. Firstly, to know about 'What is Atmosphere?' and what are its various characteristics and how those parameters affect its equilibrium. Since Physics was one of the complementary courses that we studied for undergraduate degree, it was an easier task. Also, the study of atmospheric gravity waves was carried out. Atmospheric gravity waves are generated in fluid medium or interface between two media, when the force of gravity tries to restore equilibrium. The best example of gravity waves are the ripples spreading when a stone is thrown to a river. Gravity waves has many features such as they give interesting shapes to clouds. The upward moving region of these waves is favourable for cloud development and sinking region favourable for clear skies. According to National Weather Service, Gravity waves are nothing more than a wave moving through a stable layer of atmosphere. In the atmosphere they are generated in troposphere, the lower most layer, by frontal systems or by airflow over mountains. Thunderstorm updrafts will produce gravity waves as they try to punch in to tropopause.

Secondly to study Python. Python, the programming language, created by *GUIDO VAN ROSSUM* is very simple to use and free to handle. It works based on the input or program written. The main advantage of Python compared to other programming languages is its program codes are in English language and they are very short and easy to remember. Python is a readable language due to this. Python is visually uncluttered; it has the formatting done neat and orderly arranged without any blocks in codes or statements. Python is also known as cross platform, as it works in various systems such as *Windows, Linux, Mac* etc., also has the 'the standard library' as it consists of a large collection of codes and functions to write the program by the need. When, becoming aware of 'Python', and its features, it is hard to consider python as a tough tool. The fact is that, it keeps interesting more and more, when it comes about Python programming. It is a platform, to easily organize each works with the help of supporting functions and packages that are related to kind of jobs, projects, etc.

In over the span of 25 years, Python has managed to reach a level that is high above others making it the fastest growing language. Python has also managed to dominate other programming languages such as Java, C, C++, etc. Besides all these, its object oriented feature has made it even more accurate to deal with. ("Future of Python in the IndustryNew Gen Apps"-an article by Anurag, April 13, 2020). The features of python and the whole benefits of the python programming and the demand rising because of this made it a fastest growing language. In the future, it is certain that Python will definitely be something that people can never ignore for any reasons in the industry. Because, it has gained that much importance in the field.

Python codes are useful to graph any function easily. Using it we first plotted density verses altitude graph using the corresponding formula, which shows the inverse relation between them. Secondly the pressure profile where pressure increases with decrease in altitude. The next one is the graph which plots altitude vs amplitude of atmospheric waves. Energy of atmospheric wave is also mentioned here. The gravity waves are plotted with height in y axis and amplitude in x axis and also vice versa. Properties of gravity waves are established then. Gravity waves with different values of frequency are plotted. Thus, we arrived at a conclusion that when frequency increases the wave become thinner. After that energy is varied keeping the frequency same. As energy increases, amplitude decreases, this is the inference got at the end. The atmospheric parameters such as Density, Pressure, etc. are functions of height. The exponential curves drawn indicates the inverse relationship between them. The relation between amplitude and altitude is a direct relationship, through which the vertical propagation of Gravity waves is obtained. Sine waves whose amplitude increases with increase in height is plotted which indicates the propagation of Gravity waves in vertical direction.

The following are graphs of one of the final results we obtained, it indicates the vertical propagation of gravity waves when frequency is fixed and energy keeps decreasing.

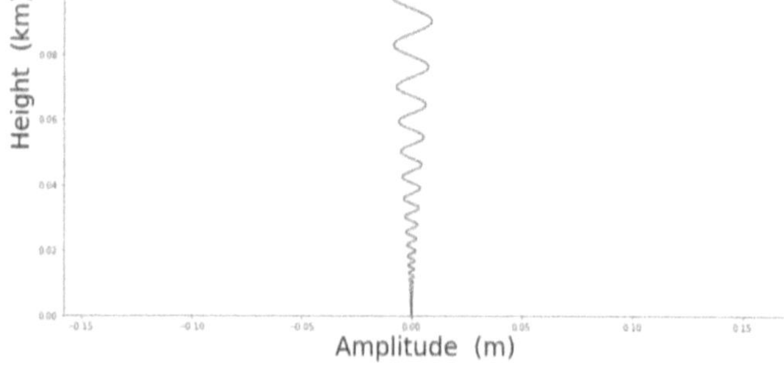

Figure 4.12: Vertical propagation of Gravity waves with Energy E=2000J

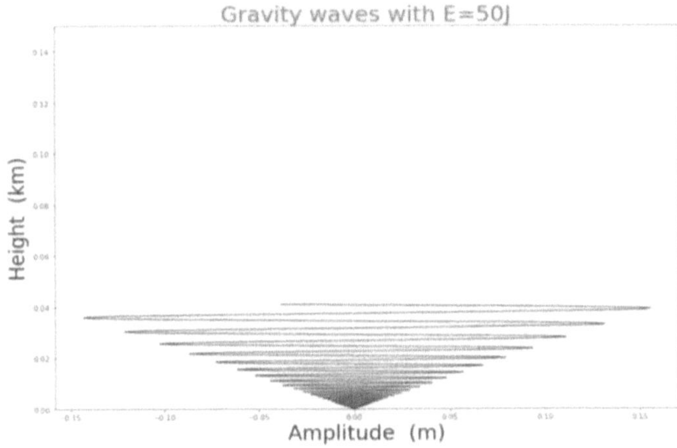

Figure 4.10: Vertical propagation of Gravity waves with Energy E=50J

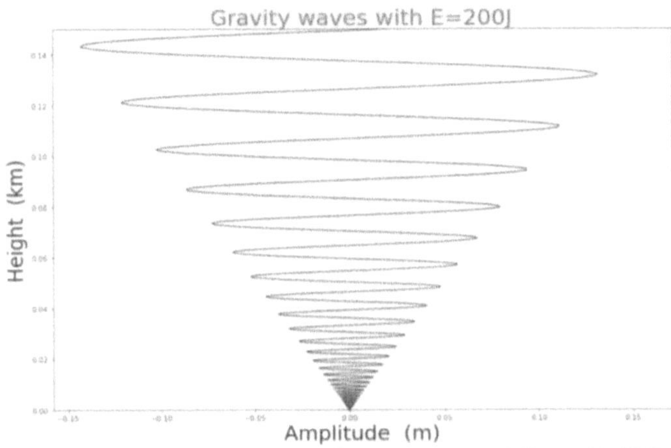

Figure 4.11: Vertical propagation of Gravity waves with Energy E=200J

There are many applications of Python in the field of Mathematics. Many mathematical computations can be done easily through simple codes of python. Expressing mathematical functions as python functions, mathematical operations on vectors, plotting of curves, vectorization of functions, solving ordinary differential equations are some of the important applications of Python. We have only made use of one of these applications, that is plotting of curves, using Matplotlib tool of Python. The infinite possibilities of python are to be explored more and this project is only the first step of it.

CLIMATE CHANGE COMMUNICATION THROUGH EDITORIAL CARTOONS: AN ANALYSIS OF SELECTED CARTOONS.

Lakshmi Prabha P
Department of Mathematics
2009-12

Climate change is a crisis of global concern. The urgency to communicate about climate change and its hazardous effects has come to be appreciated as a valid journalistic objective (O'Neil, 2017). Popular media narratives on climate change reflect the dynamics of the knowledge and power systems that it is embedded in. Media discourse is often attuned to the perceived threats and benefits of a particular climate action that can come about in groups or nationalities. The mental, temporal and spatial dimensions of climate change communication are indeed affected by the media that is put to use to convey a complex and inter-sectional concept like climate change This paper is about the cartoon visualization of climate change and how they are integral to climate change communication. Secondly I try to explore the multi-modal metaphor construction of climate change and related phenomena. My main argument is that a geopolitical and intellectually contested topic like climate change is simplified through multi-modal editorial cartoons.

Prior studies have examined climate change communication in films, art and journalistic practices. Art and Arts based practices are increasingly seen as powerful ways of developing a meaningful connection with climate change (Bentz and O'Brien, 2019; Rahul Shrivastava et al. 2012,). Within the news media climate journalists form an interpretative community and share interpretations about climate change to a considerable degree. (Nurmis, 2014 and Maxwell, 2010). According to Tobasco (et al, 2018) climate change documentaries are becoming a popular tool to reach a wide –ranging audience in efforts to elevate concern, raise the prominence of the issue in political dialogue and stimulate action. Cartoons remain an underutilized focus for research in exploring and interrogating representations of climate change(Manzo, 2012).

Editorial cartoons are a form of visual news(Greenberg. 2002) and a mode of symbolic expression (El Refaie, 2003). Cartoons employ visual metaphor as a channel for communicating the message. The symbols in them are structured by sourcing suitable codes that are prevalent in the sociocultural and political milieu in which the practice of the cartoonists unfolds and also the location of the audience. For an easy example the symbol of elephant in US editorial cartoon would mean the Republican Party and in India would mean the Bahujan Samajwadi Party. Such genre conventions are aplenty in editorial cartooning and are important markers of cartoon discourse. Cartoons reveal a side of political culture not found in official memoranda, public speeches and newspaper editorials, theoretical tracts and ideological pamphlets (Forceville, 2002).

CORPUS AND METHOD OF ANALYSIS

The empirical basis for the study is a set of 50 editorial cartoons published globally and available on the internet via a SEO search tag 'climate change and editorial cartoon'. A purposive sample of 6 was further culled from the set to describe salient multimodal metaphors that were noticed. The study of cartoons available on an online database makes research more accessible and is of pedagogic scope too. However it has to be said that since editorial cartoons are part of a newspaper which is print or digital, their online location is however removed from their discursive environment. British art historian E H Gombrich(1971) was the first to apply a systematic study of visual metaphor in caricature stating that metaphors were one of the main weapons in the cartoonist's armoury.

MULTI-MODAL METAPHORS

Lakoff and Johnson (1980) made the famous claim that metaphor is primarily a matter of thought and action and only derivative of language. A metaphor comprises a target and a source, the target is the topic and the source is that to which the comparison is made. No single definition of multi-modality exists that seems to wholly encapsulate the idea of its being. The description most suitable for comprehension and that which guides analysis in this paper is provided by Jewitt (2009)- multi-modality describes approaches that understand communication and representation to be more than language and which attend to the full range of communicative forms that people use –image, gesture, posture and so on. – and the relationship between them. A multi-modal metaphor would then be a metaphor which bears meaning through more than one mode. Consequently an editorial cartoon would be a multi-modal metaphor that transmits meaning through the verbal route and the visual route. For the purpose of this paper the cartoons chosen were multi-modal in nature as far as possible, that is, the target and source domains were exclusively and predominantly cued in two different modes, in this case the verbal and the pictorial mode. In all the fifty cartoons definite target and source domains were identifiable. Tracing the connotations have been helpful in determining what essential character of the source domain has been mapped on to the target domain

DISCUSSION

The source domain scenario can be analyzed in the following categories:

1) EVERYDAY OBJECTS 2) MYTHOLOGY 3)ILLNESS 4) ANIMAL METAPHORS 5) GEOGRAPHICAL METAPHORS 6) DISASTER METAPHORS

EVERYDAY OBJECTS- The boiling saucepan

In this cluster of cartoons climate change was communicated by connoting it with meanings that are often attributed to an everyday object.

Fig.1: A frog placed in water makes no attempt to escape

An example is given here (Fig.1) The source domain and the target domain are multi modally cued in this example making the 'global warming as a saucepan on heat' a verb o-pictorial metaphor. Additionally the frog in the cartoons connotes a situation where it is oblivious to the boiling saucepan much like many citizens who are oblivious to the detrimental effects of global warming. The connotations inferred via the reference might be 'ignorance', 'maintaining status quo' and 'capitalistic greed'.

MYTHOLOGY- NOAH'S ARK

This cluster cartoon showcases the staunchness of climate deniers and skeptics. The possible connotations mapped are 'stubbornness', 'belief' and 'hostility'. The usefulness of this metaphor stands to portray the steadfastness of the climate

Fig.2

skeptics and their reliance on unscientific explanations of denying climate change. Another metaphor that was used along with the mythological metaphor is that of 'Donald Trump as the epitome of climate change denial'.

ILLNESS METAPHORS- THE WORLD IN ITS DEATHBED

In fig.3 'Get well soon' the metaphors have an allegorical expression. The personification of the globe as an ailing person is a simple way of depicting that the entire world is in poor health due to the hazards of climate change. The depiction is also an invitation to see the environmental event as a transnational and trans-boundary phenomenon rather than the occupation of a few countries. The source 'illness' is pictorially rendered and is verbally cued as well.

ANIMAL Metaphors – The stranded polar bear

Polar bears, frogs, snails, dinos, koalas and elephants seemed to be the preferred fauna for putting across ideas of climate change. The stranded polar bear on a melting slab of ice is construed to be a specification of the melting poles scenario and the consequences on the population of polar bears. Aside from the metaphor, this situation is one that is widely discussed and debated and is a ready device to express the complexity of the event.

Fig.3

GEOGRAPHICAL METAPHORS-THE STANDING ISLAND

Fig.5 reinforces the idea of sinking islands due to sea level rising with the image of an island itself. The target is pictorially repressed by placing it on a small hillock amidst a turbulent sea. The island here also functions as a metonym

Fig.4

Fig.5

for the group of islands which are under severe threat due to rising sea levels. The possible connotation here might be that of 'severe threat'.

DISASTER METAPHORS

The disaster metaphor employed in cartoon seven is cued verbally and pictorial. The frequency of extreme weather events has given the audience familiarity with such occurrences. The depiction of a tornado or a fierce wind climatic event is a metaphor and metonym for climate change in this cartoon. Since the source domain and target domain share similar features, the connotation of 'extreme climate events' is visible and easily comprehensible.

Target Domains of climate change editorial cartoons showed how the cartoon discourse was spread around select topics. The data set (N= 50) was dominated with topics that are most suited to the western paradigm in climate change discourse. A further exploration with a greater data set might unveil more impressive varieties. However, the cartoons reviewed for the present paper are all stuck to represent a western or industrialized countries that dominated knowledge regarding climate change.

Fig.6

CONCLUSION

To wrap up, editorial cartoons are a kind of visual new discourse that employ genre specific communicative codes. In the case of climate change communication that occurs through editorial cartoons. The cuing that happens through the visual mode and the pictorial mode helps to simplify complex concepts. The simplicity that is achieved through multi-modal ways helps to situate the issue with image and rhetorical schemes that are easily identifiable and usable by the audience. In invisible and dynamic concept like climate change the visualization of concepts and issues reduce the mental distance to the concept. It tends to create a kind of visual vocabulary which will help the audience to spatialise their understanding. Finally, the exploration of the theme in this particular journalistic practice (or art form) helps to ascertain the pedagogic scope of editorial cartoons as a climate change literacy device.

BIBLIOGRAPHY

1. Bentz.G and O'Brien, K. (2019). Art for Change: Transformative learning and youth empowerment in climate change. *Elementary Science Anthology*. 7:52
2. El Refaie, E. (2003). Understanding Visual Metaphor: the case of newspaper cartoons. *Visual Communication.* 2:75-79

3. Forceville, C. (2002) The Identification of Target and Source in Pictorial Metaphors. *Journal of Pragmatics.* 34: 1-14
4. Gombrich, E.H. (1971).The Cartoonist's Armoury. *Meditations on a Hobby Horse and Other Essays on the theory of art.* London. Phaidon.
5. Greenberg, J. (2002). Framing and Temporality in political cartoons: a critical analysis of visual news discourse. *Canadian review of Sociology and Anthropology.* 39 :181- 198
6. Jewitt, Carey. (2009). An introduction to multi-modality. In Jewitt (ed) *The Routledge Handbook of Multimodal Analysis.* London. Routledge
7. Lakoff, G. and Johnson, M. (1980). *Metaphors we live by.* Chicago. University of Chicago Press.
8. Manzo, K. (2010). Imaging vulnerability: The iconography of climate change. *Area.* 42: 96-107
9. Maxwell, Boykoff (2010). Indian media representations of climate change in a threatened journalistic ecosystem. *Climate Change.* 99(1) : 17-25
10. Nurmis, J. (2014). Reporting on International Climate Negotiations: a test run for global journalism. *AEJMC Midwinter Conference* (pp. 1-16). Maryland: Aejmc.
11. O'Neil, M. S. (2017, September). *Frame Analysis in climate change communication.* Retrieved May 10, 2022, from Oxford Research Encyclopedia of Climate Science.
12. Shrivastava, P, Ivanaj, V. and Ivanaj, S. (2012). Sustainable Development and the Arts. *INTERNATIONAL Journal of Technology and Management.* 60: 21-26.

KEEP AN EYE ON POLLUTION

Inasna A.
Department of Computer Science
2020-22

ABSTRACT

Pollution is a term which even kids are aware of these days. It has become so common that almost everyone acknowledges the fact that pollution is rising continuously. The growth of industries and transport vehicles have been producing environmental issues that directly affect human's health and thus need a special monitoring system. To solve these environmental pollution issues, here we implement an IoT based pollution monitoring and control system. Where it detects the deleterious gases present in the environment. The proposed IoT technology detects co gasses during industry process, temperature and humidity using sensor technology and Raspberry pi. Each process will have a separate data monitoring and controlling mechanism. The sensors are going to be put in through a BLYNK native server mobile application which is able to show iot readings from sensors. The most important plans behind this work are to form a true time absolutely machine-controlled management mechanism exploiting IoT. Here we detect air and water pollution.

Keywords- DHT sensor, MQ sensor, Ph, Turbidity sensor, IOT, Ethernet, LCD, Raspberry pi 3.

INTRODUCTION

The monitoring and management of pollution parameters are risk challenges from industries or remote space. The intention of the research is to develop a valuable, accurate system to regulate the pollution parameters inflicting pollution and to diminish the impact of those parameters.

The implementation of the current project is to design an IoT blynk server based mostly system to scan and monitor pollution parameters and send the information to a pollution management authorized person.

IoT plays a major role in communication technology. It is an integrated communication technology in which all devices are connected to any remote location at any time. IoT uses smart digital and analog interfaces to connect and interact with embedded devices, sensors. IoT development provides numerous applications in different domains, such as safety, security, retail, consumer, home, industrial, manufacturing, utilities, energy resources, transportation, environment, healthcare, bio sciences automation and many others. IOT plays a major role in industries where a large number of devices or objects are interconnected through the internet. These devices communicate with people and things that provide the sensor data to the webpage and blynk local server. The main challenge of the project is to detect and monitor the level of harmful gases released during industry processes, humidity, temperature, pH and other factors impacting the environment.

SYSTEM SPECIFICATION

In proposed work, use of IoT to detect the industrial pollution presence of certain harmful gases. This device represents the working flow based on IOT industrial pollution monitoring and control.

Working principle of proposed work

In this proposed work we use a Raspberry pi board which collects the temperature and humidity parameters from the DHT sensor, CO_2 concentration using MG sensor and the MQ sensor senses smoke level in the atmosphere and output as a re-

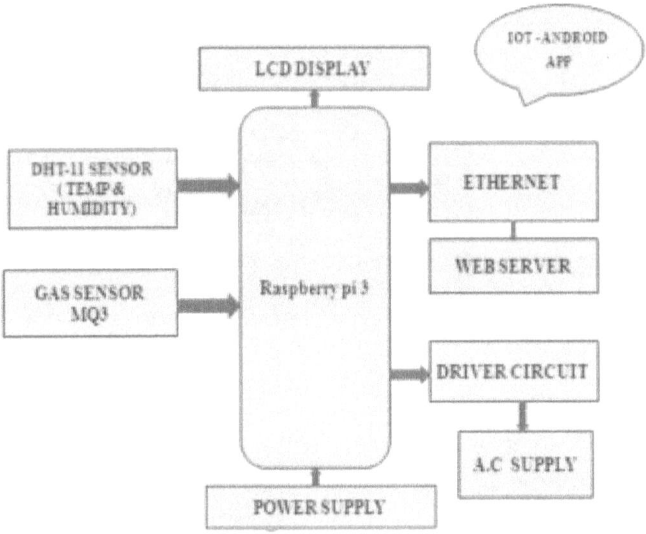

Block diagram of proposed work

sult in the form of an analog signal.

OTHER MODULES

• DHT sensor with Raspberry pi

This system will detect the harmful gases that are emitted by industries across a particular area. It simultaneously provides data to the corresponding organization. The goal of this proposed work is to display the pollution level in WebPages based mostly on harmful gas detective work and alerting systems. The toxic gases like ammonia, nitric oxide, CO_2 and various gases relating to the corresponding industries were sensed and processed by raspberry pi3. The DHT parameters are sent in the blynk mobile app for every minute. By pressing the Show Data button in the app, the iot reading and graphical data will be displayed.

DHT sensor with raspberry pi

• CO_2 Sensor

This sensor is Air pollutant sensor used for detecting pollutants in air. It measures the amount of nitric oxide and carbon dioxide. MQ sensor for measuring different kinds of gases. It detects Nitric oxide, alcohol, smoke, Ammonia and Carbon

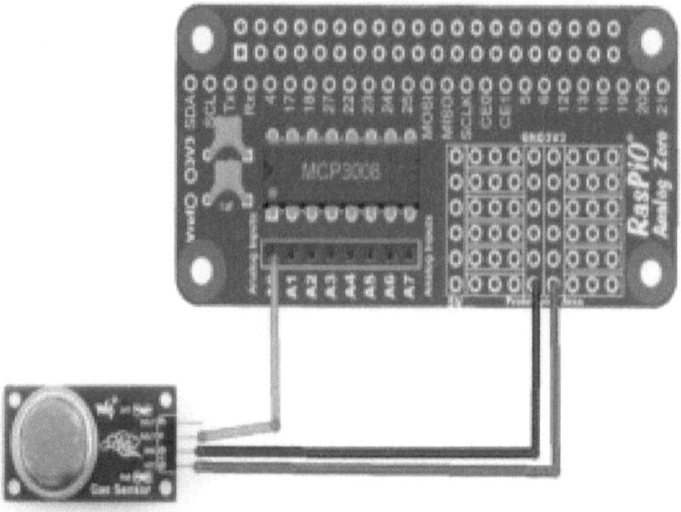

dioxide. This sensor has a low price and is particularly suitable for Air quality pollution monitors.

- **Driver Circuit**

The r driver circuit is interfaced with raspberry pi. If the pollution level is exceeded, the Raspberry pi sends a signal to driver circuit.

- **Raspberry pi**

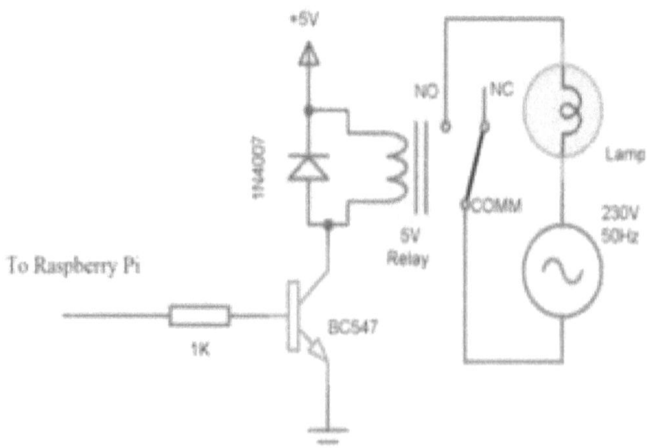

The Raspberry Pi is a single printed circuit board micro computer. The DHT and gas sensor are interfaced with raspberry pi pins. It processes the parameter and it transmits the data to the web server and BLYNK server.

RELATED WORK

In existing work they had introduced the wireless solution, based on the GSM network for the monitoring and controlling of temperature and humidity in an industrial environment. In the previous system, a pollution monitoring system was implemented by using LABVIEW software and ZigBee technology. This method is to form a system for monitoring pollution parameters and to notify the pollution control system when any of the factors crosses the threshold value declared by industry.

In the past, different methodologies and techniques were used in some of the research works.WSN plays a major role in intelligent environment monitoring; in this work they are mainly focusing on making the city environment smart, by deploying WSN all across the city, public and private transportation systems. By accessing all the dynamic global sensor networks, environmental behaviors are collected as a streaming database to identify the environmental conditions.

The architecture and prototype of the IoT system was developed. An algorithm for air pollution source estimation using Mobile Sensor Networks. In industrialization, especially the workers have been suffering from some endangering situations, so monitoring and controlling the parameters which cause pollution is necessary in the industrial environment.

CONCLUSION

The implementation of this project is to design an IoT blynk server-based system to scan and monitor pollution parameters and send the information to administration. The technology IoT primarily based atmosphere observation pollution system was designed, implemented by blynk local server. The temperature, gas, and wetness level is monitored through the net. It's helpful as compared to manual observation. It is not possible to continually watch the pollution, humidity and temperature parameters manually. The performance between accuracy and price is achieved by creating use of a single printed circuit board, digital computer Raspberry pi and sensors.

The industrial pollution monitoring and control system can further be improved by implementing sensors for controlling pollution levels thereby improving the industrial and natural environment. The local server and android app contribute to a quick response rate.

This project also detects the water pollution and pH by using a turbidity sensor. The drawback of this project is it doesn't detect the soil pollution. Only focus on air and water pollution.

OUT OF THE FRAME

Lakshmi P. M.
Department of History
2016-19

Tell us about a famous Indian photographer. Raghu rai, we most likely have the answer!!! The majority of people are aware of him. However, few individuals are aware of Homai Vyaravalla. Raghu Rai was born in 1942, while Homai Vyaravalla began her acting career in 1938. Photography, like most professions, is dominated by patriarchy. male photographers have more technical knowledge and tremendous ability, and there is an imbalance in the photographic approach. Language and imagery employed in photographic advancement, historically reinforced patriarchal control and power in the profession. Kodak cameras was the first camera firm founded in 1888, and they sold cameras using a Kodak girl. She became a well-known icon of phoney independence. For many years, there has been freedom and progress. However, the underlying meaning of words such as "even mum could use it" is significant. This is a pessimistic interpretation. It is critical to recognise the disparity in photographic experience between men and women. I'm attempting to explain some lesser-known women personalities in Indian photographic history.

In the Indian context, female photographers have always been an indisputable part of the history of photography, with a rapid increase in their participation in many genres.

Homai Vyarawalla was India's first female news photographer, capturing the British Empire's final days. Her work similarly follows the birth and development of a new country. Homai's life and work span a century in Indian history. Homai was born in 1913 into a middle-class family in Navsari, Gujarat, to members of India's small Parsi minority. Homai grew up in Bombay, where her father was a member of a touring Urdu-Parsi theatre troupe. She was the only girl in her class who finished her matriculation exam. Homai spent over three decades of her career in Delhi, where she learned photography from Maneckshaw, whom she eventually married.Homai's work is highly valuable because of her extensive knowledge. Homai's work is especially noteworthy because of her enormous photographic library, which captures both the thrill of independence and the disappointment of its unmet promises. Her survival in a male dominated sector is all the more amazing given that the profession still excludes the majority of women today. Ironically, two Western photojournalists who visited India, Henri Cartier

Bresson and Margaret Bourke-White, have received more notice than their Indian contemporaries. In an already invisible history, Homai Vyarawalla's existence as a woman was much more marginalised. In the year she received India's first National Photo Award for Lifetime Achievement, this magazine pays tribute to her unwavering energy.i

According to the material gathered to date, Maharani Monmohini, the third wife of Tripura's Maharaja Birchandra Manikya, was the first Indian woman who possessed proficiency in photography. The Maharaja was an accomplished photographer and a passionate supporter of photography's spread. Monmohini learned photography under his guidance and quickly excelled at it. She not only snapped pictures, but she also developed them. We know this because a letter written by Radharaman Ghosh, the Maharaja's secretary, was published in the Photographic Society of India's journal in May 1890.ii

Jnanadanandini Debi (1850–1941), the wife of Rabindranath

Tagore's elder brother Satyendranath, who gave a boost to women's emancipation in Bengal by her rejection of

orthodox interdictions against them, was one of the female members of the famous Tagore family who practised photography in the nineteenth century. The primary protagonists in Rabindranath's novel Jogajog are photographers: "Biprodas had a propensity for taking photographs." Kumu had also picked it up. They shot photographs at times and processed them into prints at other times." It's hardly surprising to see Tagore's Kumudini embrace photography; the writer was likely familiar with photographer Jnanadanandini and so had no qualms about putting a camera in his hands.iii

The most prominent among them are Mrs. Mayers and Mrs. Garrick. Mrs. Bibra worked at Hop, Singh & Co. with another photographer, Miss Nickles. We also know a couple more people's names. Mrs. Allen was found in Waterloo Street's Westfield & Co. in 1885. Mitter & Co., of 107 Radhabazar Street, was the first Bengali-owned studio to hire a woman. Miss S. Ballah, who worked for B. Dutta & Brothers at 367 Upper Chitpur Road, was another female photographer. Her surname reveals her Indian heritage. Sarojini Ghosh was the first professional Bengali woman to open her own studio in the nineteenth century. It's unclear whether she was the famed nationalist Aurobindo Ghosh's sister.iv

In the 1920s, a Bengali photographer named Annapurna Dutta arrived on the scene. Although she was not the first professional woman photographer, she was the first to pursue photography as a vocation and earn a living from it. She was regarded as "artist aunt" in her family, who would show up with her photography at every family gathering.v

Edna Lorenz, the most prominent name among professional women photographers in Calcutta at the turn of the century, worked for many years. Her studio was at 54 Free School Street, according to the 1918 directory. It was later relocated to 40 Park Street. Chanchalabala Dasi, a third woman professional photographer and a contemporary and competitor of Lorenz, had a studio at No. 5 Bibi Rozio Lane. Thacker Spink's Directory of 1932 lists both of their names. This route begins at one end of College Street and turns northward, passing through the Bowbazar district. The lane, which runs parallel to Premchand Baral Street, has about ten residences.vi

Debalina and Monobina Sen Roy, two sisters who worked as photographers between 1937 and 1940, were routinely featured in the Illustrated Weekly. Monobina, a lone woman photographer who had snapped the poet's image at Puri, was featured in the series "Twenty-five Portraits of Rabindranath Tagore" in 1951. Monobina Debi was married to Bimal Roy, a film director. The Photographic Society of Bengal is chaired by her sister Debalina (now Majumdar). Her photographs have appeared in several publications and magazines.vii

Around the same period, two Bengali women, inspired by their love of photography, caught moments that portrayed the inner lives of women in a country on the verge of independence - regrettably, these images were lost to history. They grew excellent at capturing brilliantly composed candid shots when lighter and portable cameras made 'untagged' photographs possible. Their grasp of highlights and shadows, as well as their compositional talents, culminated in the creation of timeless portraits. The sisters were not only photographers, but they never went anywhere without their cameras. They were ardent archivists who meticulously picked, catalogued, described, and stored their images. They published in

periodicals and shared their photographs on networks like the Postal Portfolio, allowing them to reach out to other artists and a wider audience.

Last year, Canon Philippines was chastised for having an all-male ambassador roster. In 2021, Canon USA nominated 12 women photographers out of 38 ambassadors, and 14 out of 36 this year. While Canon Europe only nominated 14 women out of 109 ambassadors, it did publish an article in which it questioned why there aren't more female photojournalists. Other leading camera brands are following a similar pattern as Canon in terms of ambassador representation. The report also discovered that people of colour had lower representation than women. The report also looked at the lack of visibility of female photographers on the social media profiles of camera brands like Canon, Sony, Nikon, Fujifilm, and GoPro. Women photographers were found to have particularly low visibility on GoPro's social media.

The Georgian version of Fast Forward: Women in Photography, organised in collaboration with University for the Creative Arts (UCA) and London College of Communication (LCC) at University of the Arts London Photography, will take place at Tbilisi Photography & Multimedia Museum (UAL). The 11th Edition of the Tbilisi Photo Festival will be highlighted by this international conference. Fast Forward: Women in Photography is a research initiative that aims to study the work and histories of female photographers, promote opportunities, and challenge dominant views in the field of photography by provoking serious, innovative conversations. The initiative, which began with a panel discussion at Tate Modern in 2014, has become noteworthy in the field of photography for scrutinising the work of women photographers and challenging the accepted canons.viii

The 4th conference, co-convened by Professor Anna Fox and Maria Kapajeva, Professor Val Williams, and the Tbilisi Photography & Multimedia Museum, Georgia, will focus on the ways in which women photographers have worked in conflict-affected nations. The meeting will contribute to fresh understanding in this understudied field. The organisers are particularly interested in bringing together and giving voice to photographers, scholars, curators, and editors from across the region, including the South Caucasus, Ukraine, Turkey, Israel, Iran, Iraq, Palestine, the Persian Gulf countries, Jordan, Afghanistan, Syria, Lebanon, Pakistan, India, and Bangladesh. Because one of the most obvious characteristics of conflict zones is the violation of women's rights, this issue resonates with many realities faced by women throughout the area, including Georgia.

Women may be underrepresented in photojournalism and high-end commercial photography, but they have cornered the market in maternity and new-born photography, a relatively new industry that exists largely because of women photographers. Women also dominate the field of new-born and child photography. These women photographers, who are often but not always moms, have the patience and perspective needed to capture the photographic style that families seek in their photo albums. Women photographers, in fact, are responsible for the look that has grown so popular in family photography. Women no longer want clean, formal images of their lives.

However, this is much more than just ensuring that women have more and better opportunities in the sector. "Men and women see life differently and have diverse views to share," the NUJ writes, "yet the idea of what constitutes 'excellent photography' has primarily been determined by men's work..." To stay current and authentic, the photography profession needs to diversify in order to accurately reflect the communities it covers." Things are gradually improving, and both women and men are opposing photography's patriarchal discourse. Les Rencontres d'Arles, the world's most prominent photography festival, takes place every year from July to September. Only 34% of the photographers invited to exhibit in 2018 were female. More than 300 high-profile creative professionals signed an open letter to artistic director Sam Stourdzé, criticising the festival's programming and urging him to aspire for gender parity in 2019, the festival's 50th anniversary. According to preliminary reports from this year's festival, such difficulties can and do have a good influence. This year's programme has many more women, including younger and rising artists as well as famous ones. Such diversity results in a more varied and complex set of visual responses to the world. Nonetheless, rather than gender equality being an intrinsic component of the festival, it needed a challenge to the established quo for the issue to be addressed. However, the type of inherent equality that the business requires is beginning to emerge at a more grassroots level. Female-only galleries, such as Finkelstein in Melbourne, Australia, and collectives like Phrase in Cardiff and Redeye in Manchester are springing up. They claim to demonstrate that greater diversity leads to a more varied and sophisticated set of visual reactions.

Society has preconceived notions and stereotypes about our women photographers and their contributions to the society stays blurred always.Photography is one of the professions that gives equal importance to skill and technical awareness. Women had less chances to gather technical skills and the profession demanded to travel to so many places, and there was always the difficulty of balancing home and work. Despite the unfavourable stereotypes, not all women were discouraged from purchasing their first camera, but the data above show the long-term influence of gendered discrimination.

The story of women in photography is not yet complete, but it is moving forward in a positive direction. In the art world, women photographers are still significantly underrepresented. Despite the fact that more women work in fashion and commercial photography, men continue to outnumber and out earn women. Women photojournalists are few and far between, and we've only just begun to consider how women and women's opinions are portrayed in photography. However, we find hints that women are shaping photography for themselves about halfway through the story. Women are once again–or still are–the record keepers of our lives, shooting crucial milestones in families' lives, exposing societal injustices that are often buried behind politics and conventions, and capturing the human tale in harsh situations. And women are figuring out how to live their lives. And women are finding creative methods to pursue their passion for photography. In some ways, we women photographers have reclaimed ground that has been lost, but we've also made some progress. Just hoping that the entire paradigm shifts.

REFERENCES

https://quod.lib.umich.edu/t/tap/7977573.0002.205/--whatever-happened-to-rehana-homai-vyarawallas photographs?rgn=main;view=fulltext

https://quod.lib.umich.edu/t/tap/7977573.0002.205/--whatever-happened-to-rehana-homai-vyarawallas photographs?rgn=main;view=fulltext

https://homegrown.co.in/article/804002/a-woman-in-19th-century-bengal-paved-the-way-for-the-first female-photographers-in-india

https://quod.lib.umich.edu/t/tap/7977573.0004.202/--zenana-studio-early-women-photographers-of bengal?rgn=-main;view=fulltext https://www.telegraphindia.com/opinion/peeping-behind-the-purdah-early women-photographers/cid/455848

https://www.alamy.com/robert-knowles-photograph-by-edna-lorenz-calcutta-image450027548.html

https://scroll.in/reel/942516/images-slice-of-life-works-by-pioneering-amateur-woman photographer-manobina-roy

https://www.tbilisiphotofestival.com/en/

MATHEMATICAL ANALYSIS OF COVID-19 EPIDEMIC MODEL

Niketha Mohan, Anita Perera & Arathi Babu
Department of Mathematics
2019-22

This paper attempts to describe the outbreak of Severe Acute Respiratory Syndrome Coronavirus 2 (COVID-19) through an epidemic model. This virus has dissimilar effects in different countries. India is now in the updated stage of COVID-19 spreading, it will be an epidemic very quickly if proper protection is not undertaken based on the database of the transmission of the disease. This paper is using the data of COVID-19 upto march 2021 for the mathematical modeling and its dynamical analysis. We bring in a new representation to appraise and manage the outbreak of infectious disease COVID-19 through SEQIR pandemic model, which is based on the supposition that the infected but undetected by testing individuals are send to quarantine for an incubation period. During the incubation period if any individual be infected by COVID-19, then that confirmed infected individuals are isolated and the necessary treatments are arranged so that they cannot taint the other residents in the community. Dynamics of the SEQIR model is presented by basic reproduction number Ro and the comprehensive stability analysis.

SEQIR Model

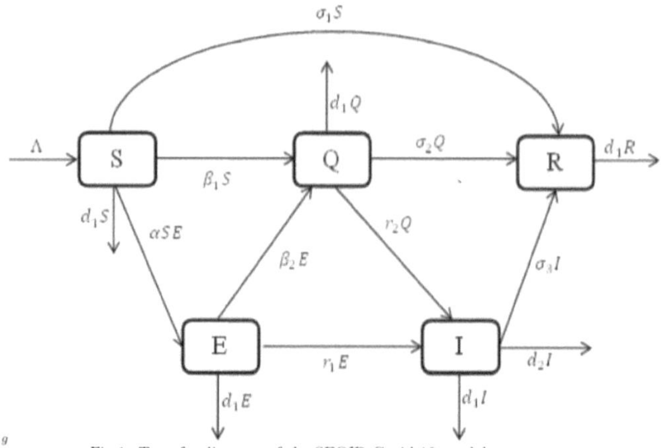

Fig.1: Transfer diagram of the SEQIR Covid-19 model

The present model structure of COVID-19 describes the dynamics of five sub-population of Indians such as susceptible(S(t)), Infected but not detected by testing (E(t)), Quarantined (Q(t)), Confirmed infected under treatment in isolation (I(t)) as well as population who live in secured zone not affected by COVID-19 virus R(t)). We assume the total Indian population, N(t) =S(t) +E(t)+Q(t)+I(t)+R(t).

$$=\Lambda-\alpha SE-\beta_1 S-\sigma_1 S-d_1 S$$
$$= \alpha SE-r_1 E- \beta_2 E- d_1 E$$
$$= \beta_1 S+ \beta_2 E- r_2 Q-\sigma_2 Q-d_1 Q$$
$$= r_1 E + r_2 Q -\sigma_3 I-d_1 I -d_2 I$$
$$= \sigma_1 S + \sigma_2 Q+ \sigma_3 I-d_1 R$$

Diseases-Free Equilibrium and the Basic Reproduction Number

The diseases-free equilibrium (DFE) of the proposed SEQIR model is acquired by setting E = 0, Q = 0, I = 0 and R = 0. The obtained DFE is E0(A, 0, 0, 0, 0).

Basic Reproduction Number (BRN) is a number which is defined as the new infective formed by a solitary infective individual for the duration of his or her effectual infectious epoch when introduced into an utterly susceptible populace at equilibrium. R0 value is always greater than zero.

SENSITIVITY ANALYSIS OF R_0

The BRN R_0 is mainly sensitive to alters in α. The value of R_0 will be enhanced if we raise the value of α. On the contrary, the value of R_0 will be as well reduced in the same proportion if we diminish the value of α. Again it is also noticed that the parameters β_1, σ_1, d_1, r_1 are related to R_0 inversely. Therefore, any increasing value of any of this mentioned parameters will definitely reduce the value of R0. But the decreasing value of R_0 will be comparatively slighter. Since the effect of parameters β_1, σ_1, d_1, r_1 and β_2 are very small on R_0, it will be sensible to focus efforts on the reduction of (transmission rate at which the susceptible individual converted to exposed individual). Therefore, the sensitive analysis of the basic reproduction number emphasized that prevention is better than treatment.

STABILITY ANALYSIS OF DISEASES -FREE EQUILIBRIUM

This section is constructed with an aim to study the stability nature of the Diseases Free Equilibrium (DFE) E0 with the assistance of subsequent Theorems:

Theorem 1 The SEQIR model structure at DFE E_0 (, 0, 0, 0, 0) is locally asymptotically stable under the condition $R_0 < 1$ and became unstable if $R_0 > 1$.

Theorem 2 The SEQIR model structure at DFE E_0 is globally asymptotically stable (GAS) under the condition $R_0 < 1$ and become unstable if $R_0 > 1$.

Existence of Endemic Equilibrium Point and Its Stability

This section reflects the existence of endemic equilibrium as well as demonstrates its stability nature to stumble on the endemic equilibrium E1 (S*, E*, Q*, I*, R*) of our wished-for SEQIR structure:

$$\frac{dS}{dt} = \frac{dE}{dt} = \frac{dQ}{dt} = \frac{dI}{dt} = \frac{dR}{dt} = 0$$

if E* has unique positive value if R0-1> 0 i.e., R0 >1. The endemic equilibrium E1 exist and stable R0 >1.

Theorem 3 The SEQIR model structure has a unique DFE E0 (, 0, 0, 0, 0) for all parameter values as well as the system has also a unique endemic equilibrium E1(S*, E*, Q*, I*, R*) under the condition R0 >1.

CONCLUSION

Our mathematical study is based on COVID-19 virus spread in India.The basic reproduction number R0 is calculated from our anticipated model. It is observed that when R0<1 then DFE E0 is globally asymptotically stable. Again, from sensitivity analysis of R0 we observe that the most sensitive parameter of our model structure is α. Also the endemic equilibrium E1 exist and is stable if R0 >1. In this study our main aim is to predict mathematically the number of infected individual due to COVID-19 situation in India. Analyzing all the results of our proposed model and observing the situation of different countries, we may conclude that India may be in a big trouble in very near future due to COVID-19 virus. To avoid this big trouble, Indian Government should take stricter measures other than quarantine, lock down etc. So far the Indian Government are continuously changing its policies to protect India from COVID-19 virus. Lastly, we say that public of India should help the Indian Government to fight against this dangerous COVID-19 as per the proposed model.

SMART HAND BAG FOR WOMEN'S SAFETY

Mrudhula E
Department of Computer Science
2020-22

ABSTRACT-

Women harassment is increasing at a rapid rate in the current scenario. Women's are not safe in many locations including workplace, while travelling. This paper presents a smart hand bag for providing security for women. This smart bag consists of Arduino board, emergency switch, voice sensor, camera and other modules. This bag works along with ~~the~~ smartphone. Using phone, messages can be send immediately to the police and emergency contacts, and can track real-time location. The system gets activated either by pressing the emergency switch or by sensing the voice. Upon pressing the switch, real time location and situation is sent to the registered contacts, mentioning the condition. An alarm gets activated and high intensity light starts flashing. Also a pepper spray starts spraying pepper.This system will allow women to travel safely.

INTRODUCTION

Smart hand bag is a safest tool for women, it can be used while travelling alone or travelling through unsafe places. The device comprises of Arduino board that has a camera for getting the visuals, a buzzer for producing continuous sound, a light system for flashing high intensity light and a pepper spray for attacking the person. Smart phone is used to track the location which is activated by pressing the emergency switch.

Systems like smart band are already used for women safety. Those devices have the ability to track location and measure pulse rate for analysing health condition. This application will also help to make emergency call to police station.

In our proposed system along with the applications of smart band, we provide a voice sensor, used to activate the device if the victim is unable to press the emergency switch ;an electric shock for hurting the attacker through the bag ;a camera for getting real time situations ;a buzzer for producing continuous sound ;a light for flashing high intensity light and pepper spray for defence.

SYSTEM OVERVIEW

This is an Arduino based system. The circuit diagram is shown in figure-A. When a woman feels unsafe either she can press the emergency switch or she can make a voice command that will be sensed by a voice sensor. Then the system will automatically activate. A smart phone is connected to the device via Bluetooth. Upon activation, an application in the phone automatically detects the location and send messages to the registered numbers and to the Women Cell.Those messages contain the situation of the victim and the exact location.It also sends real time visuals using the camera to the women cell. A pepper spray is activated. A loud voice and light are also produced to inform the neighbours.

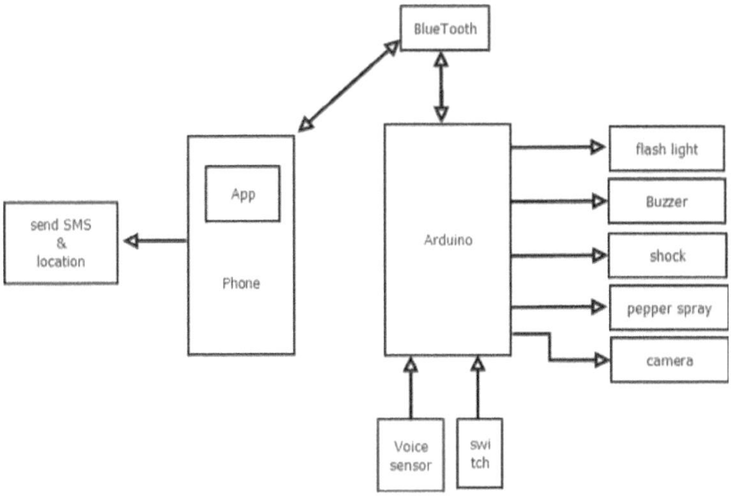

Figure-A

OTHER MODULES-

Voice sensor

The sound module is very sensitive to the environment sound intensity, commonly used to detect the intensity of the sound of the surroundings. When module, in the intensity of the sound environment, cannot reach a set threshold, the OUT will output a high level, when the intensity of the sound from the outside environment more than a set threshold, the module OUT output is at a low level; Small digital output board OUT can be directed to the microcontroller, through single-chip microcomputer to detect the high and low level, thus to detect sound environment;

- Can detect the intensity of the sound environment,
- working voltage 3.3-5V
- Output form: digital switch output (0 and 1)

Flash light

The power LED is triggered by pulses generated by Arduino and create vision obstruction to the attacker.

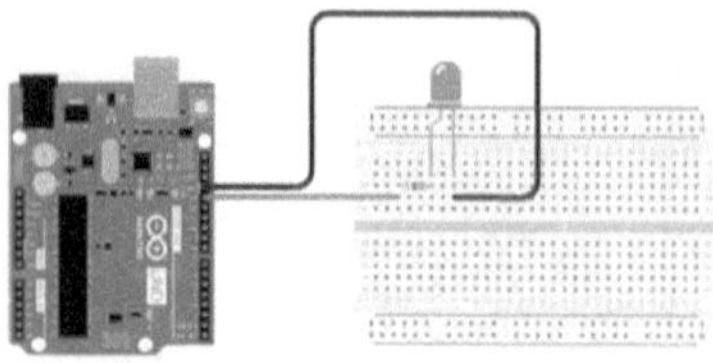

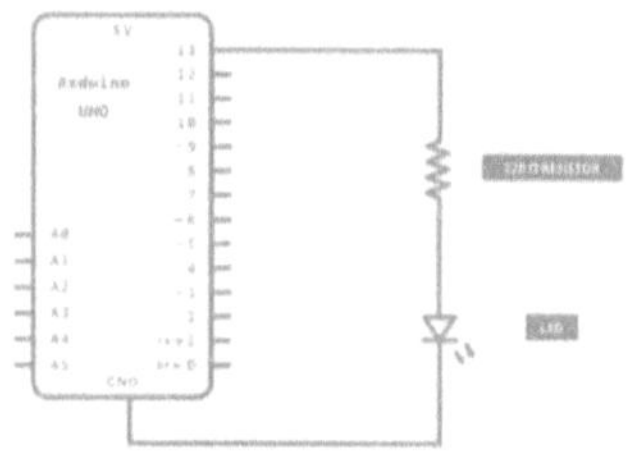

Buzzer

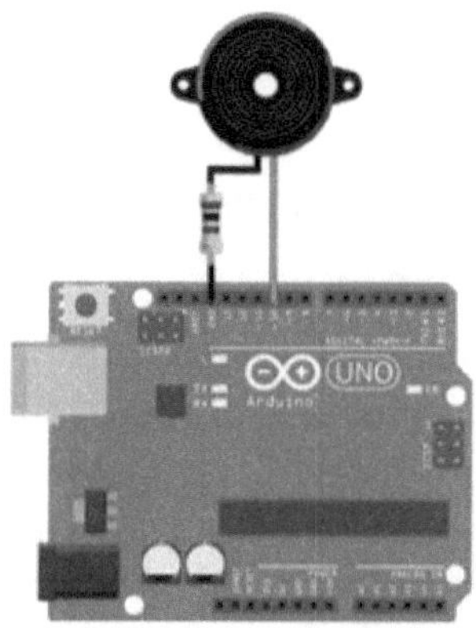

It is basically a tiny speaker that you can connect directly to an Arduino. You can make it sound a tone at a set frequency The buzzer produces sound based on reverse piezoelectric effect. This will help to produce sound that will alerts nearby people.

CONCLUSION

Women face a lot of threats and harassment in many ways. So we are presenting a smart hand bag for women safety, which can send real-time information for getting instant help from the police authority. Now, women can travel at night time and through lonely places safely.

PSYCHOLOGICAL WELL-BEING OF NURSES WORKING IN GOVERNMENT HOSPITALS AND PRIVATE HOSPITALS

Neeraja S. Mahesh
Department of Psychology
2019-22

ABSTRACT

Health workers across the globe play a vital role in building the resilience of community members. The aim of the present study is to compare the psychological well being of nurses working in government and hospitals and private hospitals. Participants of the study consisted of sixty nurses (30 nurses working in private hospitals and 30 nurses working in Government hospitals) from Wayanad and Kozhikode districts of Kerala, India. Data was collected after administering Ryff and Singer's psychological well-being scale to know participants' level of their psychological well-being. Results of an Independent sample t-test computed indicated that there is significant difference in the psychological well-being of nurses working in government hospitals and private hospitals of Kerala. When compared to the psychological well-being of nurses working in private hospitals, nurses working in government hospitals show greater psychological well- being. The study implies the need for further in-depth exploration in this area and also highlights the need for coming up with different strategies to enhance psychological well-being of nurses working in the private sector as well.

Key Words: Psychological well- being, Nurses, Government Hospitals, Private Hospitals

INTRODUCTION

Health workers are the backbone of our health sector. While the "Kerala model" was established, health workers were there to modify our hospitals, reduce child death, and for everything they are capable of (Govindan, 2000). Their selfless work is what drives our health sector forward. Health workers, mainly nurses, work for hours continuously under a lot of pressure. During covid -19, health workers worked endlessly for months, they were not even able to see their loved ones. From multi-speciality hospitals that contain a large number of health workers to primary health centers that contain a small number of health workers, nurses are the ones that mostly deal with patients (Asianet News, 2020). Their mental health has an important effect on how they work, which in turn decides the quality of health services people get (Dayma & Mohan, 2017). Psychological well being is a variable closely related to mental health. (RYFF, 1995).

Psychological well-being is simultaneously the absence of the crippling elements of the human experience–depression, anxiety, anger, fear–and the presence of enabling ones–positive emotions, meaning, healthy relationships, environmental mastery, engagement, and self- actualization. Psychological well being is above and beyond the absence of psychological ill-being and it considers a broader spectrum of constructs than what is traditionally conceived of as happiness (Seligman & Csikszentmihalyi, 2000; Seligman, 2011).

In addition to the psychological components theory of well-being, other theories explain well being in terms of social, emotional and cognitive factors. Subjective well-being theory explains well-being in terms of both affective (how we feel) and cognitive components (what we think), (Diener et al. 2015).

Studies are conducted all over the world to understand the well-being of health workers. Findings of a study conducted by Weibelzahl, Reiter, & Duden (2021) revealed that health-care professionals were most affected by protective measures at their workplace and changes in work procedures. Psychological symptoms, particularly anxiety and depression, were significantly more severe than in a non-clinical pre-pandemic sample and in the general population during the pandemic. At the same time, most professionals indicated that they would not seek help for psychological concerns.

This all highlights the need for research to explore the well-being aspects of healthcare professionals, especially nurses. Lack of studies to understand the well-being of health care workers in Kerala motivated the researcher to study more on this area. Findings of the present study will provide a baseline understanding on where our nurses, especially those working in government and private sector stands on the well being scale would help policy makers and organizations to develop different strategies to support this very important group of our society. Ensuring the psychological well-being of health workers makes them stronger to fight efficiently against any epidemics (Dayma, Mohan, 2017).

AIM

To compare the psychological well-being of nurses working in government hospitals and private hospitals of Kerala.

OBJECTIVE

To study the psychological well-being of nurses working in the government sector and private sector.

HYPOTHESIS

There will be significant differences in the psychological well-being of nurses working in the government sector and private sector.

METHOD

Participants

Participants included 60 nurses (30 nurses working in government hospitals and 30 nurses working in private hospitals) aged from 23 to 50 selected from different hospitals in Kozhikode and Wayanad districts of Kerala using random sampling technique.

MEASURES

Psychological well-being scale (Ryff and Singer, 1988)

Psychological Well-Being Scale developed by Ryff and Singer (1998) consists of 18 items rated on a 7-point scale ranging from 1 (strongly disagree) to 7 (strongly agree). The six dimensions measured by this scale are autonomy, environmental mastery, personal growth, positive relations with others, purpose in life and self-acceptance. The reliability of the tool was.80. The re-established reliability coefficient of this tool is.82.

PROCEDURE

The psychological well-being scale is a self -administering questionnaire that can be given in person to the respondents. For nurses working at the government hospitals after establishing rapport clear instructions were given as follows: "*There are 18 questions with 7 options. Read carefully each statement and put a tick mark on the option which you feel is much appropriate for you. There are no right or wrong answers. All your responses will be kept confidential and will only be used for research purposes. It will take only 10-15 minutes to complete this questionnaire. Kindly ensure that you have not missed any items. Please feel free to raise your queries*".

For nurses working at private hospitals, questionnaires were given in the form of Google form with clear instructions and their responses were recorded.

Scoring

Positive items are scored in the order 7 for strongly agree, 6 for moderately agree, 5 for slightly agree, 4 for don't know, 3 for slightly disagree, 2 for moderately disagree and 1 for strongly disagree. Negative items are scored in the order 1 for strongly agree, 2 for moderately agree, 3 for slightly agree, 4 for don't know, 5 for slightly disagree, 6 for moderately disagree, and 7 for strongly disagree.

STATISTICAL ANALYSIS

An important statistical measure used in the study is the t-test. The t-test is used to test if there is a significant difference between the mean or average score between the two groups. (Kothari, 2004). All the testing was done with the Statistical Program for social sciences (SPSS) software.

RESULT AND DISCUSSION

Preliminary Analysis

Preliminary analysis of the data was done by using descriptive statistics. Data collected from 60 nurses (58 females and 2 males) were used for the present analysis. Samples were analyzed based on the psychological well-being scale (PWB scale). This scale consists of 6 sub-factors namely autonomy, self-acceptance, positive relationship with others, purpose in life, environmental mastery and personal growth. Details are given below.

Table 1 Descriptive statistics (N=60)

Variable	Mean	Median	Mode	Standard deviation	Skewness	Kurtosis
Psychological Wellbeing	94.86	97	104	12.60	-1.11	1.86

From table 1 it is found that the mean, median and mode values for psychological well-being scale is 94.86, 97, 104 respectively. Analyzing the values, it is found that values show slight differences between each other. Whereas values of skewness and kurtosis are close to zero. Hence normality of the data across the variable is assumed and the parametric test is used for the present study.

Comparison between nurses working in the government sector and private sector

To test hypothesis-1 which states that "*there will be significant difference in the psychological well-being of nurses working in government sector and private sector*" an independent sample t-test was conducted. Here significant differences between the two groups are tested for total score of psychological well-being scale. Details are given in table 2.

Table 2 Independent sample t-test (N=60)

Group / Variable	Nurses working in government hospitals (N=30)		Nurses working in private hospitals (N=3-0)		t-Value
	Mean	SD	Mean	SD	
	98.63	9.99	91.10	13.93	2.41*

*Significant at 0.05 level

From the above table, the t value obtained in the total Score of psychological well-being scale is 2.41 which is significant at 0.05 level. It indicates that there is a significant difference in the psychological well being of nurses working in private hospitals and government hospitals. When the means of the two groups were analyzed, it is found that the mean and standard deviation for nurses working in government hospitals are 98.63 and 9.99 respectively. And for nurses working in private hospitals the mean is 91.10 and standard deviation is 13.93. From the obtained result it can be interpreted as those nurses working in government hospitals have greater psychological well-being when compared to nurses working in private hospitals. Therefore, we can accept hypothesis-I which states that "there will be significant difference between psychological well-being of nurses working in government hospitals and private hospitals' '.

Even though the researcher reviewed related research, studies pertaining to psychological well-being of nurses during the covid-19 pandemic period were found lacking. But many studies dealt with general well-being of nurses (Dayma, Mohan, 2017). From previous research, it seems that the working environment, salary, and leisure-time time play an important role in psychological well-being. Therefore, by altering these factors psychological well-being of nurses in private hospitals can be improved.

CONCLUSION

From the present study a significant difference was found in the psychological well-being of nurses working in private hospitals and government hospitals. When compared with nurses working in private hospitals, nurses working in government hospitals have greater psychological well-being.

REFERENCES

Asianet news, (2020), കോവിഡ് 19: ആരോഗ്യപ്രവർത്തകരുടെ മാനസികാരോഗ്യം; ശ്രദ്ധിക്കേണ്ട കാര്യങ്ങൾ, Asianet news, March 20.

Dayma. & Mohan. (2017). Psychological well-being of nursing staff employed in Government and private hospitals, *International Journal of Education and Management Studies,* 7 (1), 27.

Gavin, B., Hayden, J., Adamis, D., & McNicholas, F. (2020) Caring for the psychological well-being of healthcare workers in the COVID19 pandemic crisis., Royal College of Surgeons in Ireland, Journalcontribution.https://hdl.handle.net/10779/rcsi.12866114.v1

Ryff, C., &Keyes, T. (1995), The structure of psychological well-being revisited, *Journal of Personality and Social Psychology, 69.*

Ryff, C. (1995), Psychological well-being in adult life, *Current Directions in Psychological Science, 4* (4), 99-10.

Seligman & Csikszentmihalyi, (2000), Positive psychology an introduction, *American Psychologist*, *55* (1), 5-14.

Weibelzahl, S., Reiter, J., & Duden, G. (2021). Depression and anxiety in healthcare professionals during the COVID-19 pandemic. *Epidemiology and Infection*, *149*, E46. doi:10.1017/S0950268821000303

LIFE CYCLE OF THE RED PIERROT BUTTERFLY *TALICADA NYSEUS*

Naveena E
Department of Zoology
2019-22

ABSTRACT

The Red Pierrot is a small butterfly seen in the Indian subcontinent. Life cycle is completed in the host plant *Kalanchoe spp*. Breeding was continuous. Eggs laid singly on the underside of leaf. Eggs hatched in 3 days. Larval stage lasted for 13- 15 days. It consisted of 5 instars. Each instar lasted for 3- 5 days. Pupal stage lasted for 8 days. Change from egg to adult butterfly was completed in 24- 27 days.

INTRODUCTION

The Red Pierrot is a small striking butterfly seen in the Indian subcontinent and South- East Asia. They are usually found in the semi- arid plains, semi- evergreen forests, hill stations and forests. They are found wherever there host plant *Kalanchoe spp* is abundant. The caterpillar bores into the leaves and feeds on the tissues of its host plant. The adult butterfly feeds on the nectar.

HABITS

The Red Pierrot is a small butterfly that is commonly seen in India. It is a weak flier and it always remains close to the ground. It is usually active in the noon. During rest, it sits with its wings closed to show the striking patterns on its wings.

DESCRIPTION

It has a wingspan of 3 cm and has bright colors on its wings. The upper side is black except for an orange color on the lower edge of the hindwings. The lower side of the forewing is a white color with black spots and towards the base of the hindwing, it has an orange band with white spots. Males and females look morphologically similar.

HOST PLANT

The host plant of Red Pierrot is *Kalanchoe spp*. It is a succulent plant. The caterpillar burrows into the leaf and stays inside. When it feeds it leaves black markings that are filled with its waste. So, when seen from the top it looks like a rotten leaf. This serves as protection from predators. After feeding the entire leaf it comes out to pupate.

LIFE CYCLE

It is seen breeding during February – April.

EGGS: - The eggs are laid on the underside of leaves, also on the stem and on the flower. They are rounded and white. In 3 days the egg hatched.

LARVAL STAGE: - Larval stage lasted for 13-15 days. It consists of 5 instars. Each instar lasts about 3- 5 days. As the caterpillar grows its color changes from light green to pale yellow. Its body is covered by hairs. Along with its back, black dots can be seen. It spends its larval stage inside the leaf and comes out to pupate.

PUPA: - The caterpillar came out of the leaf and stopped feeding. After standing for a long time in one place it began to pupate. First, the hair was lost and then it contracted its body slowly and became pupa. Pupa was yellow. Black dots and small hairs appeared on the pupa. The pupal stage lasted for 8 days. Before eclosing the pupa became black. On the 9nth day, the butterfly emerged.

CONCLUSION

Life cycle of the Red Pierrot butterfly was studied. Each stage of its life cycle was analyzed and recorded.

THRISSUR POORAM: THE FESTIVAL OF FESTIVALS

Sanam Jahan
Department of Travel and Tourism
2019-22

Tourism is travel for pleasure or business and the theory and practice of touring, the business of attracting, accommodating and entertaining tourists and the business of operating tours. The World Tourism Organization defines tourism in terms that go beyond the common perception of tourism as being limited to holiday activity not only as people travelling to and staying in places outside their usual environment for not more than one consecutive year for leisure, business, and other purposes.

Festivals are a wonderful exposure. One gets to taste local food and wine, buy local handicrafts and clothing, and attend local carnivals and functions which mark the very spirit of the festive season packing the entire nation into a hilarious frenzy. Festival is an event ordinarily celebrated by a community and cantering on some characteristic aspect of that community and its religion or cultures. It is often marked as a local or national holiday, Mela, or Eid. A festival constitutes typical cases of glocalization, as well as the high culture-low culture interrelationship. Festivals help protect local cultural traditions, develop tourism and promote the economic, social and cultural developments of the destination. In recent years, festivals are being used for promoting tourism. The time-limited event encourages tourists to visit the place during the event. This provides them a unique chance to interact with the local community, gaining a deeper experience of the ambience, customs and local tradition

Festival tourism is exploring a region at its height of fun, amusement, cheerful crowd and colourful spirit when the entire community is out to celebrate an event or a festival that had been ingrained in the local tradition and culture dating back to the distant past. Festival tourism is a portal to vent energy through channels of fun boiling with vibrant activities. During the festive season, the locality of the country observing the festival practically throbs with activities. The activities are multifarious and go on with a feverish pitch.

Thrissur Pooram often termed as the festival of festivals in Kerala is held in Thrissur, a northern district of Kerala. The word Pooram means group or meeting. The tradition has that every year the Gods and Goddesses residing at temples in the neighbouring regions come together for a day of celebration at the famous Vadakkumnatha Temple, dedicated to Lord Shiva. This coming together is led from the front by Goddess Bhagavati and Lord Krishna, presiding deities of Paramekkavu and Thiruvambadi temples. The sprawling Thekkinkadu grounds around the Vadakkumnatha Temple become the rendezvous for gods and goddesses from adjoining temples.

Thrissur Pooram owes its genesis to the erstwhile ruler of Cochin State, Prince Rama Varma, popularly known as Sakthan Thampuran. Pooram is an annual event held on the Pooram asterism in the Malayalam month of Medam. This month roughly corresponds to the months of April and May.

Prior to the Pooram day, the premises of Paramekkavu, and Thiruvambadi temples witness rituals like, kodiyettu or the hosting of the festival flag; poora pradakshinam, which is a procession taken around the sanctum sanctorum and also para nirackkal an offering made to the presiding deities.

As part of the 5-day long pooram festivities there are exhibitions that take place in the town of Thrissur. Among them, the one that draws large crowds is the Aana Chamayam. This exhibition organized by Paramekkavu and Thiruvambadi temples would have on display the entire range of decorative items used to make-up the elephants taking part in the festival. These include ornamental forehead gear called the Nettipattom; the glittering parasols called Muthukuda; whisks or Venchamaram, made of Yak hair; the circular peacock feather fan called Aalavattom and many other items.

Early morning, on the pooram day, upa poorams or representations from adjoining temples like Kanimangalam, Karamucku, Laloor, Choorakottukara, Panamukkampally, Ayyanthole, Chembukkavu, Neythilakavu and other temples in and around Thrissur town would proceed to the Vadakkumnatha Temple. The procession from Kanimangalam Temple will be the first to reach the Vadakkumnatha Temple followed by others. They all march out after paying homage to Lord Shiva.

Soon the festivities would gain momentum with the commencement of the procession from the Thiruvambadi Temple with the elephant carrying the idol of Lord Krishna and first reaches Brahmaswom Madom before going to the Vadakkumnatha Temple.

The visit to Brahmaswom Madom, the ancestral house of a Brahmin family who helped Sakthan Thampuran, (former ruler of Thrissur) in the formative days of the pooram festival is known as Madathil Varavu. As per tradition the deities rest here for a while before proceeding to the Vadakkumnatha Temple by noon.

Around the same time, the procession from Paramekkavu Temple carrying the idol of Goddess Bhagavati, its presiding deity, reaches the Vadakkumnatha Temple. The arrival of the entourage from Paramekkavu at the western entrance of the Vadakkumnatha Temple signals the start of the famed Ilanjithara Melam - the traditional percussion ensemble, consisting of drums, horns, and cymbals, held in front of the 'Ilanji' tree inside the temple premises. It is a musical ensemble that reaches its crescendo with an elated crowd cheering its mesmeric rhythm.

And finally, by evening, 15 majestically caparisoned elephants of ThiruvambadI Temple emerge through the southern gate of the Vadakkumnatha Temple to face a line up of decorated elephants from Paramekkavu Temple. Each pachyderm bears an umbrella holder, a peacock fan carrier and a yak-tail fly whisk wielder. Before dusk sets in, the spectacular Kudamattom (exchange of parasols) would begin to the loud cheer of the spectators, who by now would have gathered in thousands. This ceremony takes place to the accompaniment of Pandimelam - the traditional orchestra. The ceremony will last beyond the sunset, finally paving the way to a dazzling display of fireworks by early morning.

Most of the pooram day attractions are staged again the next day and are known as Pakal Pooram or the pooram held during the daytime. It culminates in bidding farewell to the presiding deities of Thiruvambadi and Paramekkavu temples in front of Lord Shiva, the presiding deity of Vadakkumnatha Temple. They promise to meet next year before they part. And by evening, Kalaasakottu or the final beating of drums, signal the end of the pooram festivities, followed by Vedikettu or bursting of high-sounding crackers the traditional community-feast in which people partake to have the Poora Kanji, a specially made porridge, signals the end of pooram festivities.

Kerala is the land of never-ending festivals and Thrissur Pooram is, without doubt, the grandest of all. Thrissur Pooram is not just an occasion for Gods and Goddesses, but also for mortals who cherish every moment of the warmth and the spirit of togetherness. The objectives of the research are:

- To study possibilities of international tourism by Thrissur pooram festival
- To study potential developments in the festival shopping industry.
- To study the role of Thrissur pooram in overcoming acute unemployment issues in the district.
- To study the impacts of the covid 19 on Thrissur pooram.

Thrissur is best known for its mammoth Pooram Festival, which is the most colourful and spectacular temple festival of Kerala. Thrissur Pooram attracts large masses of devotees and spectators from all parts of the State and even outside. The legends and myths behind each festival of

Kerala are many, varied and equally interesting. It was believed that every year the dynastic gods and goddesses of the neighbouring province met together for a day of celebration. This usually happened on the pooram asterism of one of the spring months.

The findings of the research are:

- Festivals often serve to fulfil specific communal purposes.
- Thrissur pooram becomes one of the chief sources of economic booster in Thrissur.
- Diversified cultural practices and rituals make pooram one of the indispensable parts of the cultural heritage of Thrissur as well as Kerala.
- Thrissur pooram is a magnificent spectacle merging the spiritual and cultural essence of Kerala hence promoting tourists across the globe thereby establishing an initiative for international tourism.
- Thrissur pooram generates immense employment opportunities in diverse sectors for creative and successful entrepreneurship.
- Thrissur pooram was recognized during the COVID era for its smooth conduct, obliged to COVID protocols without any failure.

The suggestions of the research are:

1. While the world has turned by and large to eco-friendly alternatives like the colourful drone light shows and digital spectacle, it is uncivilised to continue fireworks, life-threatening, not only to humans but to birds and animals too.
2. Augmented Reality(AR) can be made use of in the pandemic to share the experience of pooram for people who stay at home without going out to physically see the central spectacle of the pooram.
3. To curb the stagnant traffic, the government can get in touch with the organizers of Pooram and enforce special laws on the days of celebration. Setting limits on automobile movement is a fine way to counter traffic congestion.

4. Additional measures can be brought about in accident-prone areas to reduce the number of accidents.
5. setting up additional space to accommodate more people especially women can not only deal with the issue of overcrowding but also can help promote gender equality as more females get the chance to tune in.

CONCLUSION

Kerala has gained international attention for Thrissur pooram and is becoming a popular international festival tourism destination. Tourists have identified that Kerala has an immense number of festivals that exhibit Kerala's culture and tradition. Kerala is the only probable state in India where the festival is used as mainstream to attract tourists globally.

Festivals are important cultural activities and they are very significant and indigenous to the very people who celebrate them. Thus the overall analysis indicates that Kerala had developed an image as an international festival destination through the Thrissur pooram although in the light of the pandemic.

A STUDY ON EDUCATION, EMPLOYMENT AND JOB PREFERENCES OF WOMEN IN CALICUT

Saina Arfa, Shifna KP, Suparna TK, Teena Joseph And Treesa Jacob
Department Of Economics
2018-2021

INTRODUCTION

Kerala, which ranks first among the Indian States in terms of Human Development Index (HDI) and Gender Development Index (GDI) presents a poor picture in terms of female work participation. But, the most tragic failure of development in the State is the acute unemployment and low labour force participation rates of both men and women. Reckoned in terms of all the three measures of unemployment – usual status, current weekly status, and current daily status - used by NSS, Kerala has the highest incidence of unemployment both for males and females and in rural as well as urban areas. It is often held that education improves the employment opportunities of a woman and raises her earning potential. But the experience of Kerala suggests an inverse relationship between education and female employment.

In Kerala, the pace of social development has far exceeded that of economic growth leading to a paradoxical situation. On the one hand, it has encouraged women to participate in economic activities. On the other hand, structural changes in the economy have put severe constraints on the prospects of female employment at the higher levels and led to a process of casualization of the female labour force.

The study is a modest attempt to explain female labour force participation in the State and all relationships that affect women and their efforts.

About women and employment we can say that, the number of entries on Indian women's employment increases every year and offers a range of specialized information.

OBJECTIVES

The study is an empirical investigation into the identification of factors that determine female labour force participation in the State. It tries to explain female labour force participation in terms of inter-linkages between education, employment and job preferences. Specifically the objectives are;

1. To identify the impact of mothers education on their employment.
2. To find out the impact of mother's education on daughter's employment.
3. To study the relationship between female work participation and level of education.
4. To identify the role of females in financial responsibility, activities performed at home and their ownership on assets.

IMPORTANCE OF THE STUDY

The economic development has improved the lot of a few nations over the past decade or two. But at the same time, the period has also witnessed degradation in the lives of large numbers of people especially in the developing countries. And what is more surprising is the fact that economic development has shown a gender bias, i.e. as high as 70 percent of the world's poor, even today, are women. Besides, the socio-economic differentials between men and women continue to increase even in countries of the West; and the unemployment situation is worse for women than for men in both developed and developing countries. So for an economy to develop we need development from both men and women. So the study took the part of women to know about the factors that hinders a women from growing.

METHODOLOGY

The methodology includes the source of data, sample size and tools used for data analysis

A) Sources of Data

The study makes use of both primary data and secondary data.

Primary Data : The primary data were collected from females of age between 24-65 years. It is obtained with the help of a questionnaire filled by 50 females from Kozhikode district. We also collected information on 10 females with the help of Google Form. In total we have collected data of 60 females.

Secondary Data : The secondary data were obtained from various books, reliable documents and from websites.

B) Sample Size

The study was done among 50 different sample households in the study area. Also 10 more samples were collected through Google Form.

C) Tools for Data Analysis

The tools used are simple mathematical tools like sum and percentage. Graphical representations are also made.

REVIEW OF LITERATURE

In the book "Gender, Education and Work: Determinants of Women's Employment in Kerala"(2008), Alice and K Navaneetham studies about the women's education and employment in the specific case of Kerala. One of their hypothesis is that, the economic growth and education increases the female work participation. In Kerala where women are more educated, experience the lowest female work participation and highest unemployment. The work participation among women in Kerala is 23 percent which is one of the lowest in India. The low labour force participation leads to high unemployment. There are rural and urban areas in Kerala. According to NSSO data, the female work participation rate in rural area is 22 percent. In urban area, the female work participation rate is slightly higher than the rural areas. One of the strongest reason that prevents a women entering the work force is household responsibilities. The decision of a women to participate in labour force is the outcome of the decision making of the household to which she belongs. Some of the other factors that determine women's entry into the work force are: education, age, marital status, economic status, place of residence and husbands employment. It is said that education improves the employment opportunities of a women and raise her earning potential. But in Kerala we can see an inverse relationship between education and female employment.

About the job preference of women, we can see a similarity in the job preference of educated women and an uneducated women. It seems that educated females have a strong preference for white collar and salaried job and they prefer not to take up manual work. However, the uneducated females also have strong job preferences based on factors such as social status, social hierarchy and class consciousness.

PROFILE OF STUDY AREA

Vengeri is a village in Kozhikode block in Kozhikode district of Kerala, India. It belongs to North Kerala division. It is located 12KM towards North of district headquarters Kozhikode. 10KM from Kozhikode town. 391KM from State capital Thiruvananthapuram.

PROFILE OF RESPONDENT

A. Age Group of Respondents

SL.NO	AGE GROUP	TOTAL NUMBERS	PERCENTAGE
1	20-30	8	13
2	30-40	14	23
3	40-50	26	43
4	50-60	10	17
5	60-70	2	4
	Total	60	100

Source: Sample survey 2020-21

The table shows the age of the respondents. Here we can see that more respondent are belonging to the age group of 40-50 with 43 percent of the total respondents. Respondents of the age group 60-70 were the least with just 4 percent. Hence we can conclude that our collected data is relevant to an extent because majority of the respondents are between 20-50 categories.

B. Education Details of the Respondent

SL. NO	EDUCATION	TOTAL NUMBERS	PERCENTAGE
1	Uneducated	1	2
2	1-9	8	13
3	SSLC	5	9
4	Plus Two	14	23
5	Degree	7	12
6	Graduation	8	13
7	Post-Graduation	12	20
8	Professional Course	3	5
9	Doctorate	2	3
10	Total	60	100

Source: Sample survey 2020-21

The table reveals that, only 9 percent of females have completed SSLC. Among this 20 percent of respondents have completed post-graduation course. In our study area only 2 percent of females are uneducated.

C. Occupation Details of the Respondent

SL.NO	OCCUPATION	TOTAL NUMBERS	PERCENTAGE
1	Employed	24	40
2	Unemployed	36	60
3	Total	60	100

Source: Sample survey 2020-21

The table shows the occupation details of the respondent. From the above table we can find that most of the respondents are unemployed with 60 percent. Only 40 percent of respondents are employed.

Data analysis

This section aims at objective wise analysis of the collected data.Here we analyse the monthly income of the respondents, the activities performed by the respondent, their financial decisions, the final decision made by the respondent in the household and so on.

A. Mothers education and Mothers job

SL.NO	EDUCATED MOTHERS (PERCENTAGE)	MOTHERS WHO ARE STILL WORKING (PERCENTAGE)
1	48	12

Source : Sample survey 2020-21

Majority of respondent mothers were high school graduates which contributes a 38 percent of the total data analyzed. Among this the least number of respondent mothers were Degree graduates which contributes a 3 percent of the total data analyzed. Among the 60 samples, we analyzed that the majority of respondents' mothers are not working with a 88 percent of the total data analyzed. Among this only 12 percent of respondent mothers are still working.

B. Education Status of Mother and Respondent

SL. NO	EDUCATION	MOTHER TOTAL NUMBER (Percentage)	RESPONDENT TOTAL NUMBER (Percentage)
1	Not Educated	4 (7)	1 (2)
2	Lower Primary	5 (9)	0 (0)
3	Upper Primary	18 (30)	7 (12)
4	High School	23 (38)	6 (10)
5	Higher Secondary	4 (7)	14 (23)
7	Degree	1 (3)	15 (25)
8	Professional Course	5 (8)	17 (28)

The above table is a comparison table about the education qualification of respondent and respondents mother. From the above table we can see that only 2 percent of respondents are uneducated. Whereas in the case of respondents mothers it is 7 percent. A major number of respondent mothers are high school graduates with 38 percent of total data analyzed. In the case of respondent's education, a major number of respondents have acquired graduation in professional courses with 28 percent of total data analyzed. From this we can understand that people are giving importance to women education as we can see a rising pattern in the case of respondent's education and a downward pattern in the case of respondent mothers' education.

C. Employment Status of Mother and Respondent

SL.NO	OCCUPATION	MOTHER (PERCENTAGE)	RESPONDENT (PERCENTAGE)
1	Employed	8 (12)	24 (40)
2	Unemployed	88 (52)	36 (60)

The table shows the relationship between the employment status of both respondents (daughters) and respondents' mothers. Here we can say that, even though both respondent and respondents' mothers are educated, the percentage of unemployed is more than employed. In the case of respondent's mother only 12 percent of females are employed and 88 percent are unemployed. In the case of respondent only 40 percent of females are employed and 60 percent of females are unemployed. In both the cases unemployed females is more than employed females. But in looking the other way, we can also say that the number of employed respondents is more than the respondents mother.

D. Respondents Education and Respondents Job

SL.NO	EDUCATED RSPONDENTS (PERCENTAGE)	RESPONDENTS WHO ARE WORKING (PERCENTAGE)
1	51 (85)	24 (40)

Source : Sample survey 2020-21

The table shows the education and job status of the respondent. In the above table we can see that among 60 respondents 85 percent of the respondents are educated. But among the 85

Figure 5.5 Employment Status of Mother and Respondent

percent only 40 percent of women are still working and bringing income to their families.

E. Activities performed at home (In Percentage)

SL. NO	ACTIVITES	SELF	SHARED BY HUSBAND	NIL
1	Taking care of children	40	50	10
2	Taking care of elders	36	58	6
3	Cooking	77	21	2
4	Cleaning	70	28	2
5	Washing	70	27	3
6	receiving guest	35	62	3

The above table shows the activities performed by respondents at home. From the above table we can understand that among 60 respondents a majority of respondents are engaged in cooking with a 77 percent. After that the respondents are engaged in the work of cleaning and washing with a 70 percent for both works. The least activity performed by the respondent herself is in receiving the guest with a 35 percent. Now looking at the next column, were the respondent do the activities with the help of husband/others, we can see that a majority of respondents with the help of others are engaged in the activity of receiving the guest with 62 percent. After that both the respondent with the help of others are engaged in the activities of taking caring of the elders and children with 58 percent and 50 percent respectively. The least activity performed by the respondent with the help of others is in cooking with only 21 percent. By looking at the multiple bar diagram we can see that most of the activities performed at home is by the respondent herself than shared by husband.

F. Financial Responsibilities (In Percentage)

SL. NO	ITEMS	SELF	HUS-BAND	SHA RED	NIL
1	Food	10	63	27	0
2	Rent	0	51	12	37
3	Utility Bills	7	67	26	0
4	Medical Bills	7	71	22	0
5	Servant's Salary	3	24	13	60
6	Donation	7	33	22	38
7	Beauty and Fitness	15	20	10	55
8	Travel and Fuel	5	68	22	5

Source : Sample survey 2020-21

The above table shows the financial responsibilities taken by both respondents and their husbands. By taking the first column we can say that a major number of respondents spend their income on beauty and fitness with a 15 percent. After that most of their income is spend on food materials with a 10 percent. From the table we can see that among 60 respondents, none of the respondents spend their income on rent. And it is a zero percentage. By taking the next column we can see that a majority of income spend by the respondents' husbands are on hospital bills with a 71 percent. After that the husbands are spending their income more on travel expenses with a 68 percent. The least amount of husband's income is spent on beauty and fitness with a 20 percent. Now looking at the third column we can see that a majority of income spent by both respondents and their husbands is on food with a 27 percent. From the above table we can also see that only few houses have servants in their home to do the chores. Even though if they have servants, the financial responsibility of paying them salary is taken by the respondent's husband with a 24 percent. Overall we can say that most of the financial responsibilities are taken by the husbands..

G. Ownership of Asset (In Percentage)

SL. NO	ASSET	SELF	HUS-BAND	SHA RED	NIL
1	Vehicle	21	57	17	5
2	Gold	85	5	7	3
3	Land	12	58	25	5
4	House	12	61	27	0
5	Additional Property	15	20	12	53
6	Business Partnership	3	13	10	74

Source : Sample survey 2020-21

The above table is about the ownership that the respondent and their husbands have on assets. From the table we can see that a major number of respondents have ownership on gold with a 85 percent of the total data analyzed. The respondent with the lowest ownership is with the business partnership with a 3 percent. By looking at the the next column which is about the ownership of asset possessed by the respondent's husband, we can see that a major number of respondents' husbands have ownership on land with a 58 percent. The husbands with the lowest ownership on asset is with gold, with only 5 percent. Now in the third column in which the ownership of asset is shared by both respondent and their husbands, we can see that a major number of respondents and their husbands have shared ownership on house with 27 percent. Only 7 percent of shared ownership is there for both the respondent and their husbands in gold. Overall we can say that most of the ownership of the asset is possessed by the respondents' husbands.

CONCLUSION

Our study on the topic "Education, employment and job preference of women in Kerala " is, by necessity, an exploratory one, aimed at identifying the determinants of female labour force participation in Kerala based on micro-level data. Active participation of women in the labour force has increased tremendously, worldwide, during the past few decades. However, macro-level data indicate that there exist wide divergences in the pattern of female labour force participation between developed and the developing countries. The level of female labour force participation in India is not only low but has remained almost stagnant over the past several decades. Moreover, there still exist large differences between the male and the female work participation rates.

The decision of a woman, unlike that of a man, to participate in the labour force is the outcome of the decision-making of the household to which she belongs. Thus, it depends upon her personal and family characteristics and other intervening variables.Participation of women in economic activities outside home has an important bearing on gender relations within the household. Several studies in developed and developing countries have shown that women's relative economic power gives her control over a variety of 'life options', including various types of decisions. Women's independent incomes, consequent upon their joining the labour force, not only improve their status, but the well-being of the whole family as well. This is because women spend more for the sustenance of the family.Several studies have reported that economic independence increases the status of women in the household in general and enhances their role in decision-making.

REFERENCE

Bhende, A. (1985). *Women Employment in Kerala*. Bombay: Himalaya publishing house. http://himpub.com

Chacko, E. (2003). *Marriage, development, and the status of women in Kerala, India.* Taylor & Francis, Ltd. https://www.jstor.org

Chandna, H. (2019). *Kerala is best performing state for education of girls, Uttar Pradesh the worst*. The Peint. https://theprint.in

Devi, L. (2002). *Education, employment, and job prefer-*

ence of women in Kerala: A micro-level case study. Dr. K. N. Nair. https://www.krpcds.org

Eapen, M, Kodoth, P. (2002). *Family structure, Women's education and work: Re-examining the high status of women in Kerala*. Centre for Development Studies, Trivandrum. https://ideas.repec.org

Mathews, E.T. (1995). *Unemployed and self-Employed: Job preference and Employment perspective* (Vol.30). Economic and Political Weekly.

Mazumdar, S, & Guruswamy, M. (2006). *Female Labour Force Participation in Kerala: Problems and Prospects*. International Institute for Population Science Mumbai, India. https://www.iipsindia.org

Sebastian, A, & K, N. (2008). *Gender, Education and Work: Determinants of Women's Employment in Kerala*. Academic Foundation.

Sharma, A, & Saha, s. (2015). *Female Employment Trends in India: A Dis aggregated Analysis*: Vol.XIII. The NEHU Journal. https://nehu.ac.in

Varma, V. (2017). *High on Literacy, Low on Workforce: Why are Kerala's Women not Making it to Work ?*. The Indian Express.

Verick, S. (2014). *Women's labour force participation in India: Why is it so low ?* ILO DWT for South Asia, New Delhi. https://www.ilo.org

ABNORMAL HETEROBLASTIC EXPRESSION IN LEAF OF *PHYLLANTHUS URINARIA* Linn.

Theertha M.D.
Department of Botany
2020-22

ABSTRACT

Plants show diversity among individual members during the ontogeny. *Phyllanthus urinaria* belongs to the family phyllanthaceae and shows heteroblastic expression in their leaf shape and size. Mainly the changes in plants influence their formandfunction. The present article deals with abnormal morphological changes observed in leaves of *Phyllanthus urinaria* Linn. from the juvenile stage to the adult stage.

INTRODUCTION

Phyllanthus urinaria is commonly used in herbal medicine and has been described to have several biological features comprising anti-virus, anti-bacterial, and anti-tumor properties. This plant is commonly called as chamber bitter, gripe weed and shatter stone. The term Heteroblastic was introduced by Karl Goebel. Heteroblastic expressions are also reported in *Arabidopsis thaliana, Zea mays.* It is defined as a crucial and rapid change in form and function that occurs over the lifespan of certain plants. Most commonly in leaf form, leaf size, phyllotaxy, internode length, anthocyanin pigmentation, rooting ability, and wood structure.

In many plants, there is a gradual change from juvenile traits, present soon after germination, to stable adult traits acquired by the time flowering commences. The change in leaf morphology is one of the main features of this heteroblastic development. Leaf shape is the most evident trait that transposes during heteroblastic development. In *Phyllanthus urinaria*, apart from size changes, there is progressive transformations in leaf shape from node to node. The change in leaf features is one of the main components of this heteroblastic development. The present study involves changes in morphological parameters of leaves of *Phyllanthus urinaria*. The observable changes in leaf shape and size, the appearance of phyllotaxy, or the arrangement of leaves of seedling and adult plants have been considered for this study. The heteroblastic expression may be altered by environmental changes. Here, the focus is on the abnormal heteroblastic behaviour of *Phyllanthus urinaria* in comparison with normal heteroblastic behaviour.

REVIEW OF LITERATURE

Goebel (1898), stated that morphological changes are frequently referred to as heteroblastic development in contrast to homoblastic development, in which no marked morphological changes occur, there is no sharp distinction between these two types of development since there are always obvious differences between the vegetative and reproductive organs (Allsopp., 1967).

The factors showing the variability in leaf index included the ratio of leaf length to leaf width. This is revealed that different varieties of single species or closely related species can be distinguished by differences in leaf index. Also, the leaf index reflects the morphological adaptation of leaves to a particular environment. The Physiological conditions or environmental factors can change the leaf index of an individual plant (Hirokazu Tsukaya., 2002).

Heteroblasty is present in some plant life. Changes in the juvenile stage through heteroblastic development attain the adult stage. The article deals with morphological changes observed in leaves of *Phyllanthus urinaria* Linn. from the juvenile stage to the adult stage. This is reported that heteroblasty shown in *Phyllanthus urinaria* as well as for the genus *Phyllanthus* (Avinash Mundhra et.al., 2009).

Almost all plants show a certain degree of variation among individual members during ontogeny. Named "heteroblastic" by Karl Goebel, but was concluded that the functional significance of this type of ontogenetic change, the evolutionary trajectories in different plant groups, and even their frequency in the plant kingdom are still unresolved issues (Gerhard Zotz et.al., 2011).

The genus *Phyllanthus* is a member of the family Phyllanthaceae based on congruent plastid and nuclear DNA sequence data that have recovered well-resolved and strongly supported clades (Awomukwu et.al., 2015).

Heteroblasy shows rapid changes in plant morphology. Shoot apical meristem contained a sequence of embryonic leaves at different stages of development. Also a dormant structure, a resting bud is demonstrative of the selective pressure on plants in temperate environments only allow to grow when environmental conditions are favourable (Schoonder woerd et.al., 2022).

MATERIALS AND METHOD

For making a comparison between normal plants with abnormal ones, first, goforthe heteroblastic expression in normal *Phyllanthus urinaria* plants. Seeds of *Phyllanthus urinaria* Linn. were collected from their habitats of different parts of the locality Cherukkad, Kozhikode. The size and leaf shape in *Phyllanthus urinaria* from their juvenile to adult transition was observed. The changes from the normal pattern of leaves of *Phyllanthus urinaria* in their seedling stages was also noticed.

RESULT AND DISCUSSION

Phyllanthus urinaria shows the transition from juvenile to adult stage in their leafsize and shape. Commonly the leaf has:

1. Cotyledons,
2. Eight juvenile leaves (eophylls)
3. Leaves on lateral branches
4. Mature leaves

The cotyledons showed oblong shape with a round apex and obtuse base. The cotyledons are followed by juvenile leaves termed eophylls. These eophylls are developed up to the 8th node and with mucronate to the rounded apex andacute, asymmetric base. The first lateral branch emerges from the seventh node with obovate leaves. The subsequent branches bear mature leaves of oblong shape. Mature leaves showing distinctly mucronate tip and obtuse, asymmetric oblique base.

Successful plant production depends on rapid germination of seeds and propagation of vegetative propagules and their growth. Mature seeds of most angiosperms passthrough dormant periods. Most Successful germination requires the right conditions flight, water, and temperature. It usually begins

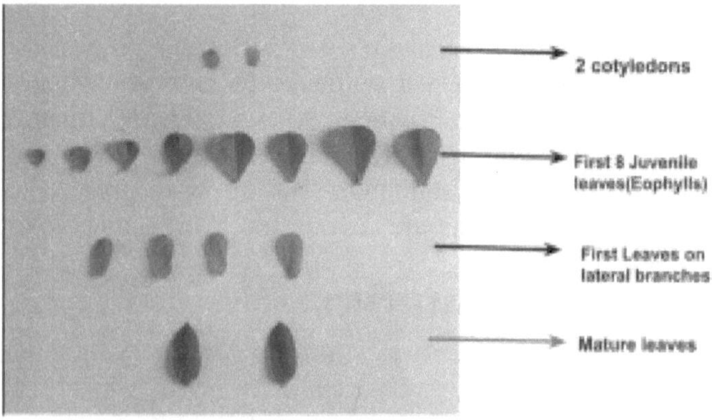

with the imbibition of water and subsequent release from dormancy. The fully developed seeds contain an embryo and, in most plant species, a store of food reserves, wrapped in a seed coat. Seeds Germinate when soil moisture and temperature conditions are favourable for them to grow.

1A and 1B describes the developmental stages of leaves, started from the first embryonic leaf cotyledons, followed by eophylls up to six in number then first leaves on lateral branches and mature leaves. Instead of the usual form of leaves, the last picture shows the development of two sets of eophylls.

C shows the unusual heteroblasty, after two cotyledons the branches split into two develops into two sets of eophylls. It was observed that after developing these eophyll 4-5 days lat-

1A

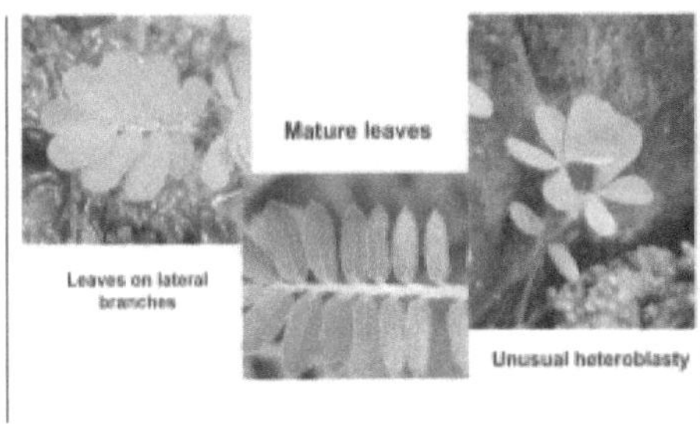

1B 1C

er the plant gets dried off. This is the reported heteroblastic development in *Phyllanthus urinaria* as well as for the genus *Phyllanthus*. This heteroblastic development help in the identification of this taxon, from other closely related species of the genus *Phyllanthus*, at the juvenile stage. The heteroblastic development in *Phyllanthus urinaria* can be of importance in morphological and physiological studies. The reference suggests that physiological conditions or environmental factors can change the leaf index of an individual plant. This was stated that in Angiosperm, not only changes in leaf index but also variations in leaf size can vary, to variations in the number of cells in the leaf lamina.

Normal heteroblasty and the next picture showing the unusual heteroblasty observed in *Phyllanthus urinaria*.

REFERENCE

Allsopp, A. (1967). Heteroblastic development in vascular plants. Advances In Morphogenesis, 6, 127-171.

Awomukwu, D. A., Nyananyo, B. L., Uka, C. J., & Okeke, C. U. (2015). Identification Of

the genus *Phyllanthus* (Family Phyllanthaceae) in Southern Nigeria Using Comparative systematic morphological and anatomical studies of the vegetativeorgans.Journal of Plant Sciences, 3 (3), 137-149.

Jones, C. S. (1999). An essay on juvenility, phase change, and Heteroblasty inseedplants.International Journal of Plant Sciences, 160 (S6), S105-S111.

Mundhra, A., & Paria, N. D. (2009). Heteroblastic expression in leaves of *Phyllanthus urinari*a Linn.Science pub, 1(2).

Schoonderwoerd, K. M. (2022). Interspecific variation in resting bud morphology and seasonal heteroblastic leaf sequences among temperate, woody plants(Doctoral dissertation).

Tsukaya, H. (2002). The leaf index: Heteroblasty, natural variation, and the genetic control of polar processes of leaf expansion.Plant and Cell Physiology, 43 (4), 372- 378.

Zotz, G., Wilhelm, K., & Becker, A. (2011). Heteroblasty—a review.The Botanical Review, 77 (2), 109-151.

A STUDY ON THE BIODIVERSITY OF MOTHS IN CALICUT

Vandana P
Department of Zoology
2020-22

ABSTRACT

Moths are a group of insects belonging to the order Lepidoptera under the phylum Arthropoda. It is estimated that there are around 1, 60, 000 species of moths in the world. Most members are nocturnal. Moths have ecological significance because they are important elements in the food chain. Some moths are pests of crop plants. Moths are indicators of a healthy environment; thus, their diversity and population can be regarded as pollution indicators. They also play a vital role in pollination and nutrient recycling. This study aims to identify the different species of moths observed in the locality and to appreciate their diversity.

INTRODUCTION:

The majority of the species under the order Lepidoptera are moths. The term Lepidoptera has a Greek origin and it means; scale wings. Moths are phytophagous, and their larva feeds on the leaves of Angiosperms. It is believed that moths coevolved with angiosperms. Moth's life cycle is composed of four stages namely, egg, caterpillar, pupa, and adult. Caterpillar is the active feeding stage of moths. Moths are significant in an ecosystem, certain members like Cotton bollworms and Corn bores are serious pests. They act as night pollinators and are potential bio-indicators. The classic example of Pepper Moth explains how changes in the environmental conditions affect the evolution of moths. After the industrial revolution, the population of black pepper moths (*Biston betularia* L.) increased. Certain species of adult moths belonging to the genus Atlas, Luna, Polyphemus, and Promethea do not feed since they lack mouthparts. They depend on the food stored during the caterpillar stage. Some moths suck nectar. The domestic silk moth, (*Bombyx mori*) is cultivated for manufacturing silk. Silkworms feed on the leaves of *Morus alba*. It is farmed for the silk with which it builds the cocoon. Moths are the prey for nocturnal insectivores like bats, owls, and certain birds. Moths are associated with Baculoviruses, immature larval forms of moths are common hosts of the virus. Thus, they can be used as effective biocontrol agents. The ecological role of moths necessitates serious research in this field.

METHODOLOGY:

The surroundings were observed to identify moth species. All observations in this article were made during the period from January 2021 to May 2022. Photographs of the moths were taken in their natural habitat without harming them. No specimens were collected or killed. The photographs were used to identify the moths. Most specimens were identified up to species level and some up to genus level

IDENTIFICATION:

The details of the observations made are summarized in the table given below.

S. No	Species	family	Author
1	*Asota caricae*	Erebidae	Fabricius, 1775
2	*Achaea janata*	Erebidae	Linnaeus, 1758
3	*Iridopsis**	Geometridae	Warren, 1894
4	*Maxates calaina*	Geometridae	Moore, 1887
5	*Mocis undata*	Erebidae	Fabricius, 1775
6	*Bastilla joviana*	Noctuidae	Stoll, 1782
7	*Palpita**	Crambidae	Hubner, 1808
8	*Dysphania percota*	Geometridae	C.Swinhoe, 1891
9	*Naxa textilis*	Geometridae	Preyer, 1884
10	*Schizura ipomoeae*	Notodontidae	Doubleday, 1841
11	*Nyctemera coleta*	Erebidae	Stoll, 1782
12	*Cryptothelea nigrita*	Psychidae	Boisduval, 1828
13	*Eupterote undata*	Eupterotidae	Blanchard, 1844
14	*Nepita conferata*	Erebidae	Walker, 1854
15	*Eudocima hymermnestra*	Erebidae	Cramer, 1780
16	*Thyas coronata*	Erebidae	Fabricius, 1775
17	*Olena mendosa*	Erebidae	Hubner, 1823
18	*Euproctis lunata*	Erebidae	Walker, 1855
19	*Eudocima phalonia*	Erebidae	Linnaeus, 1763
20	*Endoclita malabaricus*	Hepialidae	Moore, 1879
21	*Mocis latipes*	Erebidae	Guenee, 1852
22	*Caenurgia erechtea*	Erebidae	Cramer, 1780
23	*Erionata thrax*	Hesperiidae	Linnaeus, 1767
24	*Phyllodes verhuelli*	Erebidae	Vollenhoven, 1858
25	*Aporandria specularia*	Geometridae	Warren, 1894

26	*Attacus atlas*	Saturniidae	Linnaeus, 1758
27	*Pergesa acteus*	Sphingidae	Cramer, 1779
28	*Erebus ephesperis*	Erebidae	Hubner, 1827
29	*Cyana peregrina*	Erebidae	Walker, 1854
30	*Acherontia lachesis*	Sphingidae	Fabricius, 1798

*Organism identified only up to genus level.

The photographs of different species of moths are presented according to the serial order mentioned in the above table.

13

14

15

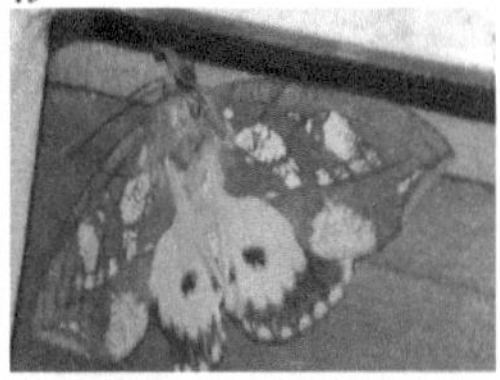

16

17

18

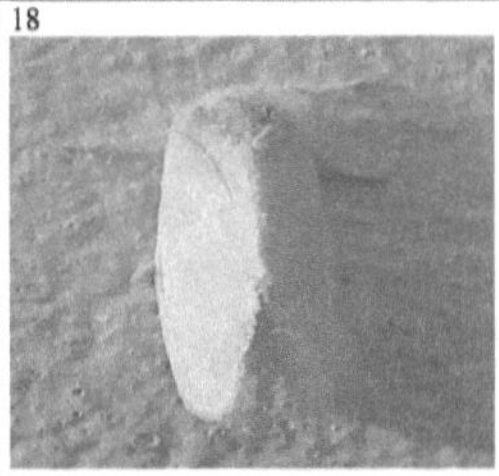

19

20

21

22

23

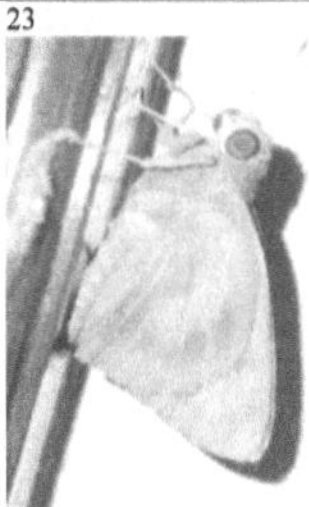

24

25

26**

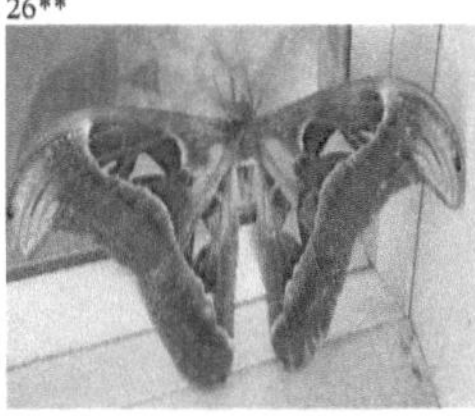

27**

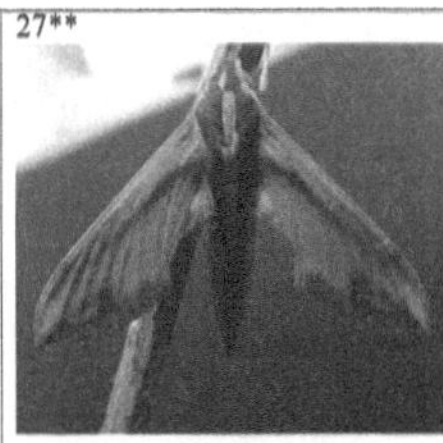

28

29

30

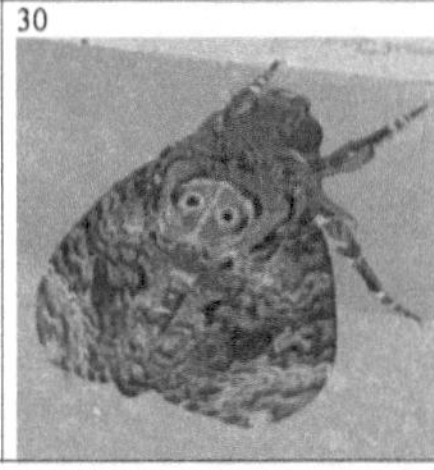

Acknowledgements

The author is grateful to the support rendered by Ancy Mariya Alex and Ajanya KK for sharing the photographs of the following moth species.

*Photo of *Dysphania percota*, was taken by Ancy Mariya Alex of II DC Zoology.

**Photo of *Attacus atlas* and *Pergesa acteus* was taken by Ajanya KK of II DC Zoology.

CONCLUSION:

Twenty-eight species of moths were identified. Two samples were identified up to the genus level. Half of them belonged to the family Erebidae. Five specimens belonged to the family Geometridae.

REFERENCES:

https://www.britannica.com/animal/moth-insect

https://butterfly-conservation.org/moths/why-moths-matter

https://en.wikipedia.org/wiki/Peppered_moth_evolution

https://www.wildlifepreserves.org/species-guide/why-moths

http://biodiversitylab.ncbs.res.in/media/SondhiEtal_MothsOfSchendurneyAndPonmudi_2018_TropLepRes.pdf

Trogon- Newsletter of Malabar Natural History Society, Edition- Jan-Apr 2020; A Preliminary Checklist of the Moths of Kerala.https://www.sciencedirect.com/science/article/pii/S1319562X21004782

https://www.mothsofindia.org/

https://www.inaturalist.org/projects/moths-of-kerala

A STUDY OF STRESS ON E-LEARNING AMONG COLLEGE STUDENTS

Wafa Sayid Ali
Dapartment Of Business Administration
2019-2022

INTRODUCTION

E-learning was introduced worldwide since the beginning of the twenty-first century in the paradigm of education technology, withholding the traditional teaching approach, upholding an often accessible, flexible, and personalised learning platform for the learners. The sustainability of e-learning with the advancement of modern technology got prioritised in the education system at the university level. A learning system based on formalised teaching but with the help of electronic resources is known as E-learning. While teaching can be based in or out of the classrooms, the use of computers and the Internet forms the major component of E-learning. E-learning can also be termed as a network enabled transfer of skills and knowledge, and the delivery of education is made to many recipients at the same or different times. E-learning is known as electronic learning, online learning, computer-based learning, or digital learning, which provides learning support using digital devices.

The deadly surge of the COVID-19 pandemic forced the students to stay at home in every corner of the earth and maintain social distancing. Thus, e-learning was prioritised by all sorts of academic institutions. E-learning seemed to be an alternative way of continuing teaching-learning activities in the current crisis. However, the arguments issued with the rapid switch of e-learning worldwide during the pandemic were compelling and whether it was free of barriers and adverse mental health outcomes like perceived stress, anxiety, depression, and students' satisfaction.

In the past 20 years, there has been considerable research concerning online learning in higher educational institutions, where full courses to even full bachelor and masters' programs have started to be delivered through online platforms. However, students who are enrolled in full instruction-based programs and are not familiar with such experience were faced with a system shock.

There are plenty of factors that explain how university students can face specific difficulties and limitations through e-learning and thus placing them in a stressful learning experience. University students are characteristically susceptible to developing stress disorder and depression and the possibilities of such implications are expected to grow. Such students will lack vital incentives for their progression in their education or career. Social isolation and reduced activity can intensify procrastination and feelings of worthlessness. The conditions may exacerbate anxiety and depression understandably. The screen produces an intense isolation that makes it challenging for many individuals to engage in back-and-forth conversation, so it is almost hard to have constructive input without looking like you're talking through a vacuum. Financial barriers can also impede their access to technologies that shall be used for them to keep up with their learning online. Therefore, this study aims at evaluating the stress symptomatology among college students.

OBJECTIVE OF THE STUDY

1. To find out the different ways of stress on e-learning.
2. To study the factors affecting stress in e-learning.

3. To find out the consequences of stress in e-learning.
4. To know about the remedies of stress due to e-learning.

STATEMENT OF THE PROBLEM

Nowadays learning can be done through electronic as well as traditional means. E-learning has transformed the educational sector by enabling students to share information and data in a relatively easier way. But this has caused a lot of stress among college students. This study examines the ways of stress, factors, remedies, and consequences of stress faced by college students.

RESEARCH METHODOLOGY

The study examines the stress faced by college students due to e-learning. It investigates the ways of stress, factors, consequences and remedies of stress due to e-learning. For this study, descriptive research design is used. The data was collected from both primary and secondary sources. The primary data was collected through a questionnaire. A Convenient sampling method was used. The sample size was 57. Percentage analysis and comparative analysis was used and the data was presented using tables and graphs.

FINDING

- From the study, the various ways leading to stress are increased workload, loneliness and isolation, to live up to the expectation of teachers and parents, accessibility problems, fear of lower performance and physical strain.
- From these, physical strain was ranked the highest and increased workload was ranked 2nd as the ways leading to stress among the respondents.
- The various factors affecting stress due to e-learning are inability to socialize, increased fear of failure, technical issues, lack of knowledge regarding technology and communication gap. From the study all the factors were perceived as contributing to stress.
- E-learning was not accessible to all the students due to range issues, inability to access laptops etc, and unavailability of network connection.

- Lack of knowledge regarding the usage of smartphones and apps used in e-learning was another reason.
- In traditional learning, teachers can give immediate face to face feedback. Students who are experiencing problems in the curriculum can resolve them quickly and directly. E-learning on the other hand still tends to struggle with student feedback due to communication gaps.
- The various consequences of stress due to e-learning are poor concentration, sleeping difficulties, depression and anxiety, difference in level of performance, greater chance of malpractice, lack of communication skill development, loss of grade, fatigue, and health issues. Majority of the respondents agree to all these as a consequence of stress due to e-learning.
- For many students, having classes online makes them easily distracted. Social media, texting, television, and family can easily distract the students and lead to less concentration.
- The overuse of the phone and due to lack of exposure to sunlight causes sleeping difficulties.
- E- learning also causes depression and anxiety among students.
- In traditional learning, exams are conducted in large rooms and in the presence of teachers to ensure that the students couldn't communicate or cheat. But e-learning cheating and doing malpractice is easier.
- E-learning also leads to inability of students to communicate properly with others.
- With the change to e-learning, students weren't comfortable with the system they were using, or it took time for them to adapt to the new system especially when it involved technology.
- E-learning can lead to fatigue. A sense of exhaustion, anxiety, stress, and burnout the students feel due to constant online classes and exams.
- With increased screen time, eyesight issues, headaches and strain have increased.

- The various remedies of stress are set smart phone usage boundaries, time management strategy, rebalance home and work life, stay social, get enough sleep, take break, make time for exercise, get organized, team activities, connect with instructor, connect with classmates, get psychological counselling, and connect with technical support, accessibility, and other department. From the study setting up smart phone usage boundaries was considered as the best remedy by the respondents followed by staying social.

CONCLUSION

The study was conducted to examine the stress of e-learning among college students. The introduction of e-learning has brought many challenges and stress. This study investigates the ways, factors, consequences, and remedies of stress due to e-learning.

The study showed that students in general have experienced stress. The students' mental and physical health are affected. It has been seen that physical strain and increased workload, were major ways leading to stress due to e-learning. All factors like increased workload, loneliness, and isolation, to live up to the expectation of parents and teachers, accessibility problems, fear of lower performance and physical strain were perceived as leading to stress. Among the consequences, all of them were the aftereffects of stress due to e-learning. From the study the best way to manage stress was to set smart phone usage boundaries and by staying social.

A STUDY ON IMPACT OF WORKFORCE DIVERSITY ON EMPLOYEES' PERFORMANCE WITH SIGNIFICANCE TO ONE MEEM GLOBAL VENTURE, SAUDI ARABIA

Zulfa Abdulla
Department of Business Administration
2019 -2022

INTRODUCTION

A diverse workforce simply means a company has built a workforce that includes many different types of people. Workforce diversity means similarities and differences among employees in terms of age, race, cultural background, religion, gender etc. These are the individual characteristics that makes an individual unique. No two human beings are same. Workforce diversity represents both a challenge and an opportunity for business. A growing number of progressive organizations are realizing the need for diversity in the workforce.

The characteristics that describe the workforce of an organization may be categorized under two groups, namely, primary dimensions and secondary dimensions. Primary dimensions are also known as internal dimensions. These include aspects of diversity over which, typically, we have no control. These dimensions make up the essence of who we are as human beings. It include characteristics such as age, gender, race, ethnicity, sexual orientation, and physical abilities etc. Secondary dimensions represent those elements that we learn or over which we have some control throughout our lives, such as education, marital status, religious beliefs, language, etc. There are various benefits to having a diverse workforce such as Increased productivity, Increased creativity, Improved cultural awareness etc. It is essential for the management of organization to recognize workforce diversity and manage it to the advantage of the organization. If this is not done, there may be some ill consequences such as Increased turnover, High discrimination complaints, Ineffective communication etc.

The purpose of this research is to study the workforce diversity in the organization and how the diversity affect the performance of the employees. In this study the focus is on the effect of age, ethnic and educational diversity on employees' performance.

OBJECTIVES OF THE STUDY

1. To determine the relationship between age diversity and employee performance
2. To determine the relationship between ethnic diversity and employee performance
3. To determine the relationship between educational background diversity and employee performance

STATEMENT OF THE PROBLEM

In today's world, workforce diversity has become an essential concern for all organizations, which creates an opportunity as well as a challenge. If the diverse workforce is managed in a very efficient and effective manner, then the issues arising out of it is handled smoothly, this will give a positive impact on the employees' as well as organization's performance. Therefore the relationship between the diversity and the performance of employees have to be studied. This study examines the relationship between age, ethnic and educational background diversity and the employees' performance with significance to *One Meem Global Venture* in Saudi Arabia.

RESEARCH METHODOLOGY

The main objective of this research is to analyze the effect of workforce diversity on employee performance in One Meem Global Venture. For this the descriptive research design was used. The data was collected from both Primary as well as Secondary data. The sample was collected using Convenience sampling method and the sample size was 50. The primary data was collected with the help of Questionnaire. Percentage analysis was used as data analysis tool and tables, pie graphs and bar graphs were used as presentation tools.

RESEARCH DISCUSSION

The purpose of this research is to examine the relationship between three diversity parameters, viz. age, ethnic background and educational background of workforce diversity and employee's performance. The target respondents were employees and managers of the One Meem Global Venture in Saudi Arabia. A total of 50 questionnaires were distributed.

Two separate questionnaires were prepared and distributed to 25 employees and 25 managers respectively. The questionnaire given to employees provided data regarding age diversity and ethnic diversity while the questionnaire given to managers gave data regarding educational diversity and about the employees performance. The data was analysed using percentage analysis and was presented with the help of tables and graphs.

FINDINGS

- Multiple voices, perspectives, and personalities bouncing off one another can give rise to out-of-the-box thinking. The businesses can reap the biggest benefits out of diversity in the workplace. One Meem Global Venture hires employees from various ethnic backgrounds and they provide opportunity for growth and development to all the employees irrespective of their ethnicity.
- The ethnic differences can cause conflicts and groupism among the employees but it was seen that in this company the ethnic differences hasn't led to much conflicts and there is no discrimination among employees due to their ethnic differences. The employees were found to be ethnically harmonious.

- Ethnic diversity has many benefits such as increased innovation, improved company reputation etc. Great percentage of employees do think that ethnic diversity plays a positive role in this organization.
- Majority of the respondents agree that ethnic differences do not affect their performance.
- Age diversity gives the company the power of rich experienced professionals as well as the young people's mindset and their brilliant work style but if not properly managed it can lead to rise in conflicts and tension. From the data analysis it is seen that the Age differences is not a reason for conflict among the employees but almost half of the employees do feel that there is lack of bonding with employees of different ages.
- Age diversity caters to good decision making skills as it will have a group of people with different age groups. In our study we can see that the company does not consider age differences when listening to the opinions of the employees.
- Career development increases employee motivation and productivity and this will adversely lead to increase in organization's productivity. From the data it is understood that the company does not consider age diversity when providing career developmental opportunities.
- The employees are recruited by considering their educational backgrounds and thus it plays an important role in recruitment. But the differences in educational background doesn't lead to conflicts among the employees.
- Even though there is no much conflicts due to the educational differences, the employees do feel lack of confidence due to this differences.
- Unequal treatment of employees lead to less productivity, high turnover and many more problems. From the survey it is understood that the company treats all employees equally irrespective of their differences in educational backgrounds.
- When it comes to the performance of the employees, the main determinant is the timeliness, followed by quality of work, aptitude and finally the quantity of work.

- Most of the managers uses motivation as the tool for improving employee performances. Incentives, training and leadership also do help to improve the employee performances.
- The managers feel that the age group of 20-29 shows the highest performance while 40-49 shows comparatively low. But they completely agree that all employees are committed to company's mission and direction.
- Majority of the managers are satisfied with the performance of the employees.

CONCLUSION

Diversity is beneficial to both employees and the organization. Though both are interdependent, respecting and valuing each other and the differences can increase productivity. In this global economy having a well-managed diverse workforce is a norm not a choice, for competing with rivals and achieving the objectives of the organization.

Now, it's quite difficult to find any organization where there is no diversity. This study was focused on finding the relationship of Ethnicity, Age and educational background diversity and the employees' performance in One Meem Global Venture in Saudi Arabia. The findings revealed that ethnic diversity does not have much effect on the employees' performance, while age diversity and educational diversity do affect the employees' performance.

Different customs, culture, language, religion does not influence the performance of employees in the organization. According to perception of the respondents the recruitment process and getting a specific position in the organization is strictly according to set norms and education background, which may help in improving employee performance. Workforce diversity leads to synergistic performance when members are able to understand and appreciate each other and capitalize on one another's experiences, knowledge and perspectives. Proper communication among employees from different diversity can easily solve the problem which arise from diversity. When one employee communicate with other and appreciate one another any kind of conflict can be solved to a great extent.

MORPHOLOGICAL IMAGE PROCESSING

Archana V P, Arya Manohar A & Chaithra P
Department of Mathemetics
2019-22

Morphological Image Processing is an important tool in Digital Image Processing. The word morphology in image analysis is used as a tool for extracting image components that are useful for the representation and description of images. It works on the basis of set theory. The main component of morphological techniques is a structuring element, which is a binary matrix of small size. It is used to modify the shape and size of an image. It can also be represented as a set of pixels on a grid assuming the values 0 and 1.

MATHEMATICAL MORPHOLOGY

Mathematical morphology is one of the methods that provides enhancement to the images.

BINARY MORPHOLOGY

Binary morphology deals with the operations on sets. In a binary image, we refer foreground pixels variously as black or 1 and we refer background pixels variously as white or 0. In binary morphology, dilation smears the foreground while erosion thins the foreground. These two operations are dual. The structuring element (SE) in binary morphology contains only the values 0 and 1.

Dilation: Dilation causes objects to dilate or grow in size.

Dilation of A by a SE B is $A \oplus B = \cup_{(b \in B)} A_b$; $A_b = \{ a + b / a \in A , b \in B\}$

Example

The binary image A is subjected to dilation with the structuring element B.

A = {(1,1),(1,2),(1,3),(2,1),(2,2),(2,3),(3,1),(3,2),(3,3)}

Set A B = {(-1,0),(0,-1),(0,0),(0,1),(1,0)}

	0	1	2	3	4
0	0	0	0	0	0
1	0	1	1	1	0
2	0	1	1	1	0
3	0	1	1	1	0
4	0	0	0	0	0

	−1	*0*	1
−1	0	1	0
0	1	1	1
1	0	1	0

A⊕B = {(0,1),(0,2),(0,3),(1,0),(1,1),(1,2),(1,3),(1,4),(2,0),(2,1),(2,2),(2,3),(2,4),(3,0),
(3,1),(3,2),(3,3),(3,4),(4,1),(4,2),(4,3)}

	0	1	2	3	4
0	0	1	1	1	0
1	1	1	1	1	1
2	1	1	1	1	1
3	1	1	1	1	1
4	0	1	1	1	0

Erosion: Erosion causes objects to shrink.

Erosion of the binary image A by the SE B is $A \ominus B = \cap_{(b \in B)} A_{-b}$, $A_{-b} = \{ a - b / a \in A , b \in B\}$

Example

The binary image A is subjected to erosion with the SE B.

A = {(1,1),(1,2),(1,3),(2,1),(2,2),(2,3),(3,1),(3,2),(3,3)}

Set A B = {(-1,0),(0,-1),(0,0),(0,1),(1,0)}

	0	1	2	3	4
0	0	0	0	0	0
1	0	1	1	1	0
2	0	1	1	1	0
3	0	1	1	1	0
4	0	0	0	0	0

	−1	*0*	1
−1	0	1	0
0	1	1	1
1	0	1	0

A Θ B = {(2,2)}

	0	1	2	3	4
0	0	0	0	0	0
1	0	0	0	0	0
2	0	0	1	0	0
3	0	0	0	0	0
4	0	0	0	0	0

Opening: Opening removes small objects from the foreground of an image, placing them in the background. It is used to find specific shape in an image.

Opening of set A by SE B is $A \circ B$, is $A \circ B = (A \ominus B) \oplus B$.

Example

The opening image is $(A \ominus B) \oplus B$

$(A \ominus B) \oplus B = \cup_{(b \in B)} (A \ominus B)_b = \{(1,2),(2,1),(2,2),(2,3),(3,2)\}$

$A \circ B = \{(1,2),(2,1),(2,2),(2,3),(3,2)\}$

	0	1	2	3	4
0	0	0	0	0	0
1	0	0	1	0	0
2	0	1	1	1	0
3	0	0	1	0	0
4	0	0	0	0	0

Closing: Closing fuses narrow breaks and long thin gulfs. It eliminates small holes and fill gaps in contours. It join the objects.

Closing of a set A by SE B is $A \bullet B = (A \oplus B) \ominus B$.

Example

The closing image is $(A \oplus B) \ominus B = \cap_{(b \in B)} (A \oplus B)$

$A \bullet B = \{(1,1),(1,2),(1,3),(2,1),(2,2),(2,3),(3,1),(3,2),(3,3)\}$

	0	1	2	3	4
0	0	0	0	0	0
1	0	1	1	1	0
2	0	1	1	1	0
3	0	1	1	1	0
4	0	0	0	0	0

GRAYSCALE MORPHOLOGY

Grayscale morphology is a generalization of binary morphology with binary OR and AND operators replaced by arithmetic MAX and MIN operators. Procedures of grayscale morphology can also perform binary morphology but not vice versa.

Dilation

Dilation of f by a SE b at any location (x,y) is

$$[f \oplus b_N](x,y) = \max_{(s,t)\in b_N} \{f(x-s, y-t) + b_N(s,t)\}$$

Example

$(x,y)=\{(1,1),(2,1),(1,2),(2,2)\}$

$b=\{(-1,-1),(-1,0),(-1,1),(0,-1),(0,0),(0,1),(1,-1),(1,0),(1,1)\}$

f

	0	1	2	3
0	6	7	3	4
1	5	6	6	8
2	6	4	5	2
3	6	4	2	3

b

	−1	0	1
−1	1	1	1
0	1	2	1
1	1	1	1

$f \oplus b$

	8	9	
	7	9	

Erosion

Erosion of f by SE b at any location (x,y) is

$$[\, f \ominus b_N \,](x,y) = \min_{(s,t)\in b_N} \{ f(x+s, y+t) - b_N(s,t) \}$$

Example

(x,y) = {(1,1),(2,1),(1,2),(2,2) }

b = { (-1,-1),(-1,0),(-1,1),(0,-1),(0,0),(0,1),(1,-1),(1,0),(1,1)}

f

	0	1	2	3
0	6	7	3	4
1	5	6	6	8
2	6	4	5	2
3	6	4	2	3

b

	−1	0	1
−1	1	1	1
0	1	2	1
1	1	1	1

$f \ominus b$

	2	1	
	1	1	

Opening and Closing

Opening of f by b is $f \circ b = (f \ominus b) \oplus b$ and Closing of f by b is $f \bullet b = (f \oplus b) \ominus b$

BASIC MORPHOLOGICAL ALGORITHMS

Morphological algorithms are for extracting boundaries, connected components and region filling.

Boundary extraction

The boundary of a set A is $\beta(A) = A - (A \ominus B)$

Example

Given Set A with a SE B.

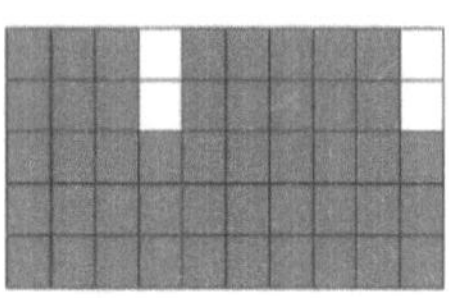

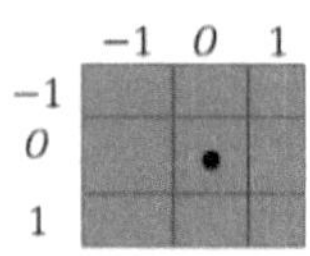

B = {(-1,-1),(-1,0),(-1,1),(0,-1),(0,0),(0,1),(1,-1),(1,0),(1,1)}

A = { (0,0),(1,0),(2,0),(4.0),(5,0),(6,0),(7,0),(8,0),(0,1),(1,1),(2,1),(4,1),(5,1),(6,1),(7,1),(8,1),(0,2),(1,2),(2,2),(3,2),(4,2),(5,2),(6,2),(7,2),(8,2),(9,2),(0,3),(1,3),(2,3),(3,3),(4,3),(5,3),(6,3),(7,3),(8,3),(9,3),(0,4),(1,4),(2,4),(3,4),(4,4),(5,4),(6,4),(7,4),(8,4),(9,4) }

The final result is the boundary given by the set difference between A and its erosion

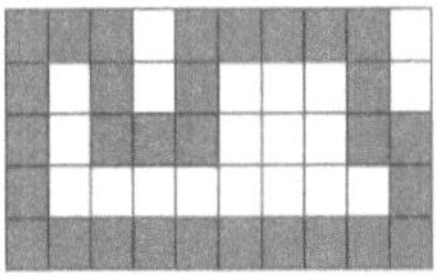

$$\beta(A) = A - (A \ominus B)$$

={ (0,0),(0,1),(0,2),(0,3),(0,4),(1,0),(1,4),

(2,0),(2,1),(2,2),(2,4),(3,2),(3,4),(4,0),

(4,1),(4,2),(4,4),(5,0),(5,4),(6,0),(6,4),

(7,0),(7,4),(8,0),(8,1),(8,2),(8,4),(9,2),

(9,3),(9,4) }

Region filling

A hole may be defined as a background region surrounded by a connected border of foreground pixels. The objective is to fill all the holes with ones. It is an iterative procedure.

$$X_K = (X_{K-1} \oplus B) \cap A^c \qquad k=1,2,3,,,,,,,$$

The algorithm terminates at the iteration step k if $x_k = x_{k-1}$. The final result is ($X_k \cup A$).

Example

Given a set A (shown shaded) with a SE B.

B = {(-1,0),(0,-1),(0,0),(0,1),(1,0),}

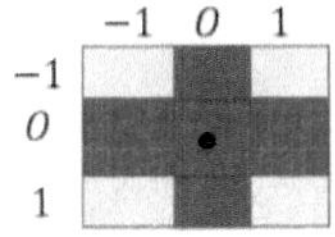

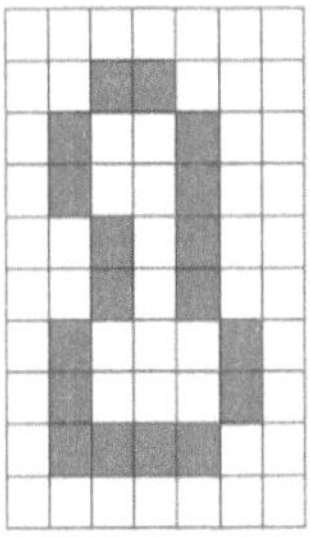

A = { (1,2),(1,3),(1,6),(1,7),(1,8),(2,1),(2,4),(2,5),(2,8),(3,1),

(3,8),(4,2),(4,3),(4,4),(4,5),(4,8),(5,6),(5,7) }

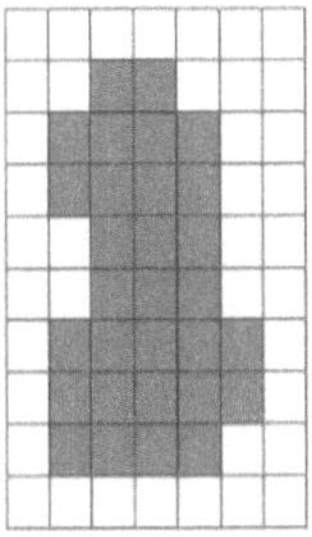

The final result is given as the union of set A and X_7.

$X_7 \cup A$ = { (1,2),(1,3),(1,6),(1,7),(1,8),(2,1),(2,4),(2,5),(2,8),(3,1),

(3,8),(4,2),(4,3), (4,4),(4,5),(4,8),(5,6),(5,7),(2,2),(3,2),

(2,3),(3,3),(3,4),(3,5),(3,6),(3,7),(2,6),(4,6),(2,7),(4,7)}

Extraction of connected components

It is used to extract a particular object from an image. The objective is to find all the connected components. It is also an iterative procedure.

$$X_k = (X_{k-1} \oplus B) \cap A \qquad K=1,2,3,,,,,,,,,,,,,,,,$$

The procedure terminates at the iteration step k when $x_k = x_{k-1}$, with x_k containg all the connected components. The final result gives all the connected components on a separate image.

Example

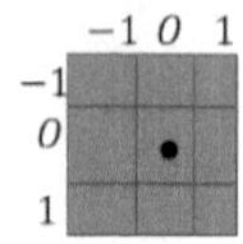

Let A be a set with SE B.

B = {(-1,-1),(-1,0),(-1,1),(0,-1),(0,0),(0,1),
(1,-1),(1,0),(1,1)}

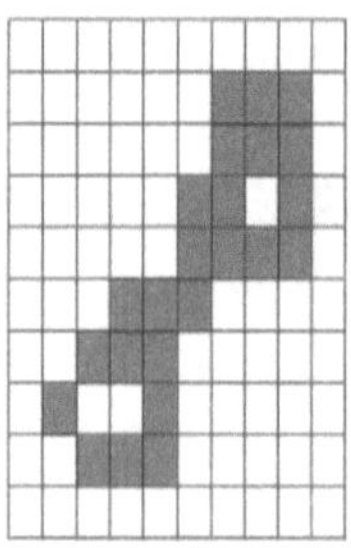

A = { (1,7),(2,6),(2,8),(3,5),(3,6),(3,8),(4,5),(4,6),(4,7),(4,8),
(5,3),(5,4),(5,5),(6,1),(6,2),(6,3),(6,4),(7,1),(7,2),(7,4),
(8,1),(8,2),(8,3),(8,4) }

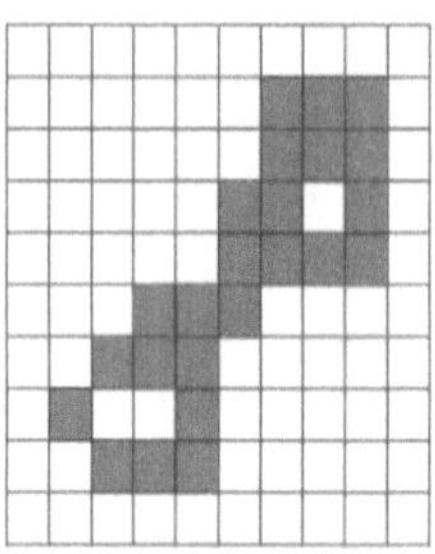

$X_5 = (X_4 \oplus B) \cap A$
= {(1,7),(2,6),(2,8),(3,5),(3,6),(3,8),
(4,5),(4,6),(4,7),(4,8),(5,3),(5,4),
(5,5),(6,1),(6,2),(6,3),(6,4),(7,1),
(7,2),(7,4),(8,1),(8,2),(8,3),(8,4)}

APPLICATIONS

Text extraction

The edges are the reliable features of texts regardless of color, intensity, layouts, orientations etc. So an edge detection scheme is applied to the input image to find all the edges. Using the edges text candidate connected components are found.

The complete process of text extraction.

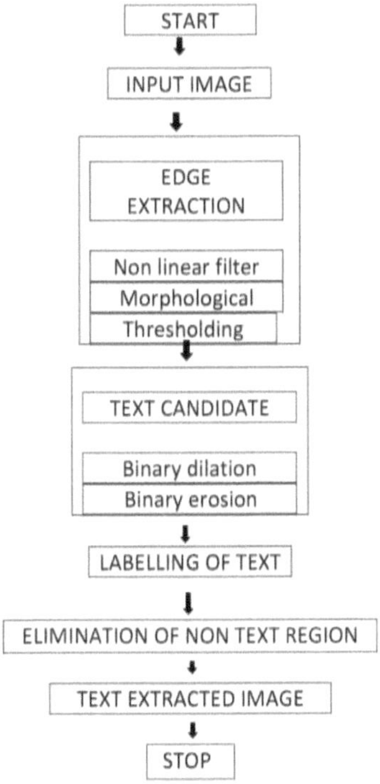

Fingerprint enhancement

Fingerprints are unique. The differences are due to the types and positions of their ridge characteristics. Hidden fingerprints are incomplete. To reduce reduction rates during matching stage, they are enhanced before matching. This can be enhanced using Morphological Image Processing.

To enhance fingerprints, they are subjected to the morphological operation.

Dilation: specific and clear ridges

Erosion: ridges are thickened

Opening: broken ridges are joined

Closing: overlapped ridges are separated

CONCLUSION

By using the morphological operators, many algorithms come in to existence and can be used in many applications like medical imaging, forensics, military applications etc. Thus Morphological Image Processing become very effective in the removal of necessary details and reconstruction of image. The purpose of this section was to provide some awareness about the operators and applications of mathematics in some of these regions.

Wind Under Our Wings

Faculty Coordinators of the R & D Cell

Mr Effy Christy Cheriyan

Dr Sr Sheeba Andrews

Dr S Sreekumari

Dr. Aswathy S

Dr Surya K

Ms Ambili Thomas

Dr. Lijiya Manju

Ms Aleena Joy

Ms Anupama P

Dr Sangeetha G Kaimal

Dr Priyadarsini P

Ms Savitha S V

Ms Smitha S

Ms Nimisha.K

Ms Priya K

Dr Vidya KT

Ms Aiswarya Paul

Dr Anu Jose

Dr Archana ER

Dr Jeena K

Sangeetha K.

Ms Vidya Krishna

Ms Shilka MV

Ms Ninikala K

Always a cut above...

Dr. Sr. Ashmitha A.C.

The numerous programs conducted by the R and D cell would have been nearly impossible without the strong support and guidance by our dear Principal Dr. Sr. Ashmitha who always had a positive spirit in encouraging the students and teachers to spearhead such programs. While the COVID pandemic made it highly difficult to engage students with informative activities, it was her sheer optimism that led us through successfully.

www.ingramcontent.com/pod-product-compliance
Lightning Source LLC
Chambersburg PA
CBHW031349230926
51848CB00002B/443
* 9 7 8 9 3 5 5 1 7 2 8 2 2 *